Profitable Plumbing

How to make the most money in the plumbing and heating trade

By M. Scott Gregg

1663 Liberty Drive, Suite 200
Bloomington, Indiana 47403
(800) 839-8640
www.AuthorHouse.com

© 2005 M. Scott Gregg. All Rights Reserved.

No part of this book may be reproduced, stored in a retrieval system, or transmitted by any means without the written permission of the author.

First published by AuthorHouse 01/19/05

ISBN: 1-4184-5489-3 (sc)
ISBN: 1-4184-5490-7 (dj)

Printed in the United States of America
Bloomington, Indiana

This book is printed on acid-free paper.

Thanks to my wife Jackie who puts up with my ideas and makes it possible for me to pursue them.

Special thanks to Ed Broughton of Chesterfield Technical Center who gave me my first training in plumbing and heating. His good nature and ability to work with the kids in his classes (10th and 11th Grade) gave many of us the start we needed to have a great career within this field.

Thanks also to W. R. Harris who gave me my break into the project management and estimating side of the business and to Lewis Picket who was my mentor in that role and one of the best people with whom I have ever worked. These gentlemen were responsible for my learning to solve problems first and work out the details later.

Thanks to Charlie Hunt* of Power & Heat Systems who gave me my start and trained me in the manufacturers representative side of the business.

Another special thanks to www.plbg.com for giving me the special outlet to help others from all over the country with their plumbing questions. This web site has done a lot for our industry and continues to do so every day. It increases the value and respect of our trade by helping plumbers share their knowledge with homeowners and each other.

*My friend Charlie died of cancer during the writing of this book. He is greatly missed.

This book is dedicated to the hard working men and women in the plumbing and heating trade.

It is my desire that this book will help you learn new methods to increase your profits.

"Every man who knows how to read has it in his power to magnify himself, to multiply the ways in which he exists, to make his life full, significant, and interesting."

- *Aldous Huxley*

Table of Contents

Forward xi

Chapter 1
Plumbing for Profit 1

Chapter 2
Getting the Plumber ready 11

Chapter 3
The Truck 21

Chapter 4
The business Plan 31

Chapter 5
The Plumbers Network of Customers 51

Chapter 6
Profit Boosters 67

Chapter 7
Time Savers, Gadgets and Specialty Tools 107

Chapter 8
Records and Bookkeeping 117

Chapter 9
Philanthropy 127

Closing Thoughts 131

Suggested Reading: 133

Acknowledgments 139

Forward

After many years of working in and around the plumbing field, I have seen first hand the wasted opportunity of most plumbers to make more money and do a better job for their customers. The main reason for this is simply that no one has ever taught them how to plan for and better serve their customers.

After looking around, I found that no book had been written to help these people become more successful. There are books on plumbing design, code and "How To" books for the homeowner, but none to help the beginning plumber get ahead or to help the plumbing service contractor grow a better business. The purpose of this book is to help you learn how to make your job as profitable as possible.

"An invasion of armies can be resisted, but not an idea whose time has come."
    - Victor Hugo

Our trade needs help. We have a shortage of good plumbers and it is not getting much better. There is a lot of money to be made by good plumbers who know how to serve their customers. A lot of money! If you are just getting started in the trade then this is the time to develop your good work habits. These include good sales skills that you will need to be highly profitable in this business.

If you have been in the trade forever and just are looking for some ways to boost your profits then this book is for you too. Try to resist the urge to flip to the "Profit Booster" chapter though. There is plenty along the way to help you set yourself apart from the crowd.

This book is like sales training for plumbers. (And more.) We will get into all of the details from the first time side work plumber all the way to the multi-truck fleet service company. We will deal with the best way to outfit the truck for the maximum benefit of the plumber and the customer, and show you how to make more money doing the same amount of work that you are already doing. We will also go into how you or your team present themselves and help your customers feel better about having you or your people in their home.

If you are just starting out, this book will give you the tools you need to develop a successful lifelong career whether you are working for yourself or for someone else. You will be able to support your family, put your children through college and help others while you become successful.

If you are already in business for yourself, this book will show you, in simple steps, how to increase profits on your very next job. It will help you raise your image in your customers' eyes and will get you more, high quality repeat business. I will also share with you tips on how to expand your business to include the best kind of customers in the best areas of your hometown. The kind whose checks are good, who will gladly pay top dollar for good work and who will give your name to their friends.

Throughout this book you will also find special quotations which I have gathered and which have helped form many of my philosophies about business and life in general. There are many people who have contributed to my success and who will contribute to yours. The people who left us these priceless little gems are far wiser then I am. I hope you enjoy them.

I know what you're thinking now. Oh great, another plumber/philosopher I have to hear about. No, it's not like that at all. It's just good common sense that seems to be lacking these days.

You will find more information in this book than you can possibly use right away. As you read it, highlight the things that interest you and will apply to your particular business or plan. Mark key pages with paperclips or sticky notes and refer to the book for refresher ideas from time to time. It will help you get into a frame of mind that will allow you to develop your own profit boosting ideas in addition to the ones represented here.

The focus and theme of this book is how to find your customer's needs and how to serve those needs. Figuring out how to spot a customer's need or getting him to tell you what he wants is key. We will discuss many different strategies for doing this. If you are able to help people get what they want, you will get what you want but it only really works well in that order.

If your approach causes a customer to buy a product or service they do not want or need, then it will reflect badly on you after you leave the property. This is not what we want. We want customers who are happy with the work and the price and who truly feel that they got a valuable service for their money. These are the people who pay your bills and make you a wealthy plumber.

This book will help you get started, find the best customers and keep them, get the best referrals, and make the most profit possible on every job, in a very demanding market. After all, that's why you went into plumbing in the first place.

So grab your favorite frosty beverage kick off your boots and jump on in.

A plumber goes to a job at the house of a brain surgeon where he fixes the doctor's sink and hands him a bill for $210.

The Brain Surgeon (clearly irritated) says, "That took you less than 30 minutes! That works out to over $400 an hour! I'm a brain surgeon and I don't make $400 an hour!"

The plumber scratches his head and says, "Yep, when I was a brain surgeon, I didn't make $400 an hour either."

Chapter 1
Plumbing for Profit

When you chose plumbing as your career, or if you are in the process of deciding now if it is for you, a big part of your motivation is the profit potential. Let's face it. This is your job. It is not your hobby. It is not a charity. One of the things I like to tell people about my job is that plumbing is not my hobby. I have plenty of hobbies, just ask my wife! All of my hobbies cost money. Some, quite a bit of money. Plumbing is not one of them. Did you get into the plumbing trade for all of the glory and fame that comes with being a plumber? I'll bet not. You are in this business to make money and lots of it.

All the profit you have ever wanted is right in front of you on every job. You just need to know how to get it. There are many honest ways for you to make a great profit on your projects without selling work that the customer does not need. What you need is to know how to get the information out of your customers so that you can have the opportunity to do the work. Many times this can result in you actually saving your customer a lot of other problems while you make more money on the job. More on that later.

As we go through this book, we will cover many ways to make more money on your jobs while you grow your business. None of these ways uses any practice of dishonesty or trickery. We do not get into high pressure selling tactics which will turn off your customers. Your goal is to do a good job at a fair price and at the same time maximize your profit. You want to achieve this goal by meeting as many of your customer's needs as possible during that service call.

The first thing to realize is that you are a salesman. If you do not think so, consider this. When you go on an interview for a job you are selling yourself and your service to the company. If you are on a service call you are selling your skills and your company's service as well as any parts necessary to complete the job. Whether or not you are aware of it, you are always selling. You can choose to fight it, (most do) or learn to use certain sales skills to your advantage.

I have met many plumbers who openly brag that they are not salesmen. If you are determined that you are not, then you are determined to stay broke, or at the very least you are limiting your chances at success.

You've heard the old saying to work smart, not hard right? Well in this business you are going to work hard AND smart. By doing so you will earn more money than most people ever could, doing the same kind of work. When you start raking in those wheelbarrows full of cash, just remember to lift with your knees, not with your back.

None of the money you ever make should come from any form of dishonesty to your customers. If you treat your customers dishonestly you will quickly end up with at least a bad reputation and at worse, sued! A bad reputation will cost you profit and you will have a very hard time changing it.

One thing is sure. If you build your business on honesty and integrity you will do exceedingly well in anything you do. Your customers will treat you well and pass your name around. Everyone wants to do business with honest people.

Profit Sources

Profit comes in many forms. You can make a profit from your labor rate, equipment, materials or tools. Later we will discuss other ways to increase profit and minimize direct work.

What many people do not think about is that there are only a few ways to really *increase* your profits. These are to raise your price, sell more (or higher quality) items on the job, or lower your costs.

Whether you are just starting out in the business or even if plumbing is not your major occupation, you need to determine your expenses, or costs. In determining costs, everything must be considered. It is not possible to plan or figure your profit potential if you do not know what all of your costs are. This part is mainly for the plumber who is just beginning to compete. The larger service plumbing outfits are well aware of what their costs are.

Cost will include things like the truck, tools, clothing, and insurance for vehicles as well as general liability insurance. Additional costs include gas, food while working, maintenance on vehicles, ongoing training and safety equipment. You may need specialty tools like drain machines, drain scopes or other items that you will use to do the type of work you choose in this business.

There might be other costs to consider, but your hourly rate and your material mark-up will need to be measured against all of them in order to see if you are making or losing money.

Profits come in as many different sizes as they do forms. Most profits are measured as a percentage of the item, service or job. All of them will wash out at the bottom of your balance sheet to show either a profit or loss. Obviously you can make a much higher percentage of profit on smaller items than you will on the big-ticket items, but in the end they should average out well.

Avoid passing anything along at cost. If you are walking down the street and a part you need falls from the sky and all you have to do is catch it and put it in your truck, you should be paid for catching it! Everything you touch is costing you something. You need to cover these costs and make a profit.

Getting Started

So how much does it cost to get into this business anyway? Well that depends. First, we will assume that you have already gone through your apprenticeship program successfully and obtained your Master Plumbers license. If you are going through it now, read on. You will find plenty in this book to help you plan for and be a better plumber from the start as well as make more money while you are going through your apprenticeship.

Since we are on that subject let me advise this: if you are either involved in your apprenticeship or working for someone else, take this opportunity to accumulate a great variety of experiences. Make sure you are able to work well with all materials possible in the trade. This should include not only plastics and copper, but no-hub cast iron, lead and oakum joints (even though you don't see it much anymore), stainless steel systems, Victaulic, cleaned and capped copper (for medical gas lines) acid waste piping like glass, Duriron, Enfield, Fuseal and others.

Learn how to be a pipe fitter and even do some welding. Learn how to work with well systems and trouble shoot them. Get boiler and/or steam experience if you can. Get a good knowledge of the repair and remodeling business, both residential and commercial. Even the repair of commercial plumbing fixtures and systems can be different from residential and can be very lucrative.

Every skill you develop and every material you are proficient in will make you more valuable than others. You will be able to get work (and keep it) when times are tough and when others without the knowledge you have get left behind. A plumber with a resume like this is hard to find and is sought after by companies doing many varieties of work.

You will be able to demand top dollar plus benefits. You will most likely be among the highest paid employees along the way, valued for your abilities. No, it won't happen overnight, but it will happen sooner than you think. Most plumbers just getting started have no idea how difficult it is to find well-trained, talented and hard working people in this trade.

You may decide to continue working for someone else. It is certainly easier to let someone else take all the risks associated with owning and operating a business. You can let an employer worry about providing benefits like insurance and vacation time for you and your family, retirement and other benefits. By using the skills that you will learn in this book you will quickly become one of their very best employees. You will be able to make more money working for another person than most people doing the same thing in this or another trade. In addition, you will be among the highest paid people in the field for that company, earn the highest commissions and when times get tough, you will be secure in your job with very little chance of being laid off. Companies know how hard it is to find plumbers like you and will do what it takes to keep you and keep you happy. You will become a protected asset to them. It's very nice to be in that position.

Everything that applies to people working for themselves to grow their profits also applies to those who work for others. If you want to be truly successful in your career, begin right now by treating that business as your own. It's your money, your truck, your tools and equipment, your customers and your profit. If it isn't now, it will be some day.

You will need a computer. This is where you will make a lot of your plans. It will enable you to track your customers, handle your bookkeeping, and print marketing materials. You can do research on new products and shop for many of your most frequently used supplies. It is an invaluable time saver for any office management task. With your computer you can also print up any marketing material you dream up to help your businesses grow.

None of this "I don't know a thing about computers" stuff. You know plenty. You just don't know how easy it is. If you don't have a computer buy one and get some tutoring from a neighborhood kid. Let's go though the bare minimum you will need to get started.

Licensing

We will assume you want to start legally and as well protected as possible as you begin. The first thing you will need if you are going to work for yourself is a business license. Contact your local Contractors Board to determine what level license is right for you. Most areas have different class licensing to accommodate businesses at different levels. Most of the time you can get started with a relatively inexpensive license which will allow you to make a pretty good income immediately without taking a test and/or charging you a high rate. As you become more successful, you will need to upgrade to a licensing level appropriate for the amount of business you do in a year.

For example, in Virginia a Class "C" Contractor license will qualify you for up to $7,500 per job and up to $150,000 per year in gross income. It costs you only $150 and must be renewed every two years. (Information current as of October 2003) This will cover you for a lot of upstart work. You may not even need to upgrade your license unless you want to make even more money or want to bid on government work. Other parts of the country will have different rates and you may not have to renew as often.

Do not attempt to launch a business without a license. This small investment will keep you out of trouble. If your local authorities find you doing business without a license, the money it will cost you to get out of that jam will far exceed any fee you could have paid up front and could cost you your tradesman license also!

Insurance

The next thing you will need is a good General Liability Insurance Policy. Any company that sells business insurance can give you a quote on this. Never do any work for anyone outside your family without both the license and insurance. Expect the first year to cost between $700 and $1000 for a policy, which will cover you to about

$1,000,000. This might sound like a lot, but here is how it works: If you are in a home and you cause damage to that home, you will have to pay for it out of your pocket if you are not covered. If you are sued and do not have insurance for your company the customer will sue you personally. One mistake could cost you everything!

To help you ease the pain of this investment let me tell you a story that happened to me personally. I got a call from a lady with a stone counter top in her kitchen. She wanted to have her sink replaced and to have a new faucet installed. She had bought the sink from a wholesaler's showroom and had paid top dollar for it. She wanted to know if I would put it in. She had called other plumbers and they had been less than responsive. Of course I got her on the schedule. She asked me if I would be able to replace it without damaging her stone top. "Of course", I told her. "I do this kind of thing all the time."

I got out there and found a cast iron lay-in kitchen sink as expected. I had a helper with me and spent some time carefully cutting the caulked joint between the sink and the top until I had it loose enough to get it out. I slid under the sink to push it up for the helper to remove it.

As I was pushing up from below I heard the unmistakable "crunch" of breaking stone! My heart sank. The counter had fractured in the corner in a thin spot in the stone where it was only about 4" from a joint. The caulk had only held on a little. It was enough to ruin my day. I felt absolutely awful. I was completely embarrassed but determined to find a solution to the problem. I quickly found out, through a contact I had with an installer of stone tops, that it could not be repaired. Also, since it was stone, not only would that top have to be replaced but in order for the kitchen to match, that island top and another section of top nearby would also have to be replaced. Total cost...$10,000! Now I was in total agony!

I stayed in close contact with this customer and my insurance carrier until the problem was resolved. It cost me a $250 deductible up front, out of pocket.

Here is the good part. When the new top was installed, I was the plumber who did the disconnect of the piping to remove the old top. I was also the plumber who did the reconnect. I checked with the insurance carrier to make sure all the charges associated with this claim would be reimbursed to the customer. I was assured that they would be. I also asked them if my charges for the necessary extra work would be reimbursed to my customer and they said yes.

I billed for labor right along with the top and by the time the job was finally done the insurance company had paid for the new top as well as for my bills on the job. The customer did not lose any money on my work and I actually made more money than I would have on the original project, even after the deductible was figured in.

In fact, since I stayed in such close contact with the customer and never left her with the impression that I would not make it good, they had me back to do other work. They appreciated the fact that I did not run from the mistake and that I immediately took responsibility for the problem, working hard until it was resolved. They are still my customers.

If I had not had the insurance, I would have been in a lot of trouble. I did not have $10,000 in the bank and probably would have had to take out a personal loan to get it. The moral of this story is if you are going to do ANY work for anyone other than your mother, get a good General Liability insurance policy. A close second moral is to chase your mistakes before they have a chance to chase you.

Other items

Of course you will need your basic tools, but there are a couple more items you should have ready to go when you make that first service call. Have an invoice, Invoices and business cards are discussed in detail in a later chapter. The cost of these items should be part of your business plan. Be sure to have everything you need on your vehicle to make basic repairs. We will discuss those needs soon. This is basically all it takes to safely start your own plumbing business.

Your success or failure depends on you. There are very few people who will volunteer to help you succeed, much less exceed.

(That's what this book will do.) You will have to find ways to differentiate yourself from everyone else with a truck out there and give your customers a reason to keep you as their plumber. This book will help you find ways to make the money a good plumber should make.

The suggested reading list at the back of the book lists books I have read and found helpful or inspirational and which have dramatically affected my career. Reading good books is the best way to keep you at the top of your game. Any information that inspires you will make you a better person. Keep reading regularly. It will enrich your life. If you are ready to get at it, we'll move on.

Chapter 2
Getting the Plumber ready

"Whether you think that you can or that you can't, you are usually right."
- Henry Ford

The Mind

Now that you have your business ready you need to get yourself ready. You have decided to make this your career. It's time to get your head ready to go into business and make more money. This is not your hobby this is your job. This is what you do to feed your family, plan for college, plan for retirement and other things. You are not in the charity business you are in the plumbing business. You are here to make a PROFIT. If you are not ready to accept these ideas you should continue to work for someone else.

If you are working for someone else you are there to make a profit for the company. Their purpose is not to provide you with a job no matter how much the boss likes you. As soon as the company stops being profitable you WILL lose your job. The more profit your company makes the better your chances are of getting a higher salary or hourly rate. Keep re-reading this paragraph. Your job is to make money for your company.

You see it *is* all about profit. The entire country works on profit. All aspects of every business come down to the same thing…making money.

People do business with people they like. To make more money than other plumbers you need to be the guy (or gal) people want to have back into their home. If you are a joy to be around, they are much more likely to have you back to fix a problem or give you that coveted referral than if you are just "there". Be outgoing, cheerful and respectful at all times. Make people comfortable to be around you.

When your customers are comfortable to be around you, you will be able to talk to them while you are working. This is the key to making more money. If they do not want to talk to you, there is no way you can make any more money than the original amount you would have made when you first came to their house.

A customer wants to see a cheerful person who is polite and at least reasonably clean. If you are a mess and/or have a bad way of communicating you will make a bad impression no matter how great a person or plumber you are.

So no matter what you want to look like or how you choose to "express yourself" off duty, if you want to make a lot of money tone it down for work. Customers are not impressed with your piercings, tattoos or your underwear. Pull up your pants, buy a belt, wear clothes that are fit for your work and leave all that extra jewelry at home.

Make sure you have a great attitude every day, at all times while you are working. Be outgoing and friendly to everyone around you all the time. Customers, coworkers and your boss will appreciate it and will help you get ahead faster. When you are happy at work you are more productive. Make people comfortable to be around you and you will be able to make more money. Do not be a grouch or a smart aleck.

If you are currently working for someone else and want to be paid more you need to realize right now that they are not in business to supply you with a job. You were hired to make them money. The more money you can make them, the more valuable you become to them and the more money you are likely to make.

Remember, if you are working for someone else, when times get tough, the first ones to be let go are the ones who make the least amount of money for the company. The "dead weight" goes first. Most companies will find a way to keep the best employees even during the worst of times because they know how hard they are to find. You need to be the very best employee your company has. If your company will not treat you fairly and you are (really) one of their best, find a new job and then quit.

Here is one thing you never want to do. Do not ever be the one or one of the crowd who speaks badly of the company. If you work for a company and they pay you, then work for them and keep quiet. If you have a problem with the company you should take it up with your supervisor or upper management and try to resolve it. If you cannot, you should find work someplace else.

Sometimes people just have personal conflicts or just do not like each other. It happens. If that someone is your boss you can do yourself a great favor by finding a job someplace else as quickly as you can. If it is a co-worker who is making your life miserable, ask to be transferred to another location. Do anything you need to do to stay productive. Your chances for advancement depend on you being your best at work. If you don't like it, don't do it. If you find that you are getting up every morning dreading going to work, it's time to move on.

One thing that you must do to give yourself the best chance at succeeding in this or any trade is to keep your driving record clean! A spotless driving record will give you great opportunities in a trade where driving is important. There are few things more frustrating to employers than having workers who are otherwise good at their job but have to be taxied around because they do not have a driver's license. Or worse yet, one who is always needs time off to go to court for this or that offense.

If you end up in a category like this you will be one of the ones who gets "weeded out" during the first slow period. You will become a burden to the company. They will replace you eventually and it certainly will limit your opportunities elsewhere.

Many plumbing contractors (as well as other trades) realize this and will not hire any prospective employees who do not have a clear record. Many demand a copy of your driving record at the time of your application.

If you are just starting out and you have some kind of issues with your record, take any steps you can to get it cleaned up. In many states there are driving schools in which you may enroll to add positive points.

If it is DUI's that are the problem it will be harder and take longer to clear up but you do not have a choice. You should consult your lawyer (if you have DUI's you have a lawyer) and find out how to get this part of your life cleared up permanently.

Since we are on the subject, many companies are going to do drug testing in order to reduce their liability insurance costs. If you are going to have a problem passing a drug test, you are not going to be a desirable employee to anyone. If you fail a drug test you will quickly be out of work.

If your company decides to go to drug testing after your employment and you fail it, you will be out of work quickly. There are all kinds of ways to beat a drug test, but none of them are successful. The truth is that all the drug free, well informed people you work with know what's going on. In other words, they know you. Employees find out who is trying to beat the test. You will be found out by a spot test or a more sophisticated test, but you will be found out.

If you have a problem with or use any drugs just quit. If you can't, get help. If you choose not to get help you will never succeed in this trade.

Here's the good part. If you do get your act cleaned up and then keep your past private, the people who did know about your past will forget it. Before you know it you will be succeeding like never before. You're going to be happier and have a lot more money. You'll get and keep better friends and your life will get on the right track. You will take responsibility for your own life and will become aware of opportunities you never saw before.

The Body

Your attitude is only part of the equation when making people comfortable around you. Right or wrong, people make quick judgments about you as soon as you get out of your truck. A sloppy or weird appearance can shut down your opportunities right away. You've heard the saying "Dress for success". It is as true in our industry as any other.

Choose good quality work clothing that washes well and is durable. It should be able to take all of the crawling, bending and other abuse that you will put it through and still be presentable at the end of the day.

The best work clothes ever invented are the good old overalls. They are durable, comfortable and only cost about the same as a good pair of jeans. You never have to pull up your pants and things are much less likely to get into them while you are crawling under a house. If you have the choice, I suggest that you make this part of your every day uniform. Once you try it, you'll never want to do hard work in pants again.

When working under a customer's home, consider keeping a pair of coveralls in your truck. When you finish the job, change into your clean clothes. Your customer will appreciate it when you go into the house to present the bill. Also, you will be clean when you show up for the next job. The next customer doesn't know what you've been doing all day and doesn't care. Keeping a tidy appearance is not hard most of the time, It just takes a little planning.

The next thing may seem silly to talk about but we have to mention personal hygiene. Basically, don't stink at your job! Nobody wants to be around someone with body odor. Get whatever products you need to keep this from happening to you. If you have to, keep some extra deodorant on your truck as a "refresher" so your next customers are not offended. The last thing you want is bad thoughts lingering about you once you have left your customer's home.

How about your hair? If you have hair that is just going to get messed up anyway, get a good hat to wear. If you are thinning or balding you may not think it's as big a deal but a ball cap gives a bit of bump protection and cleans up your appearance nicely. It can also serve as a conversation starter. Especially if it's sports related. Funny how something as simple as the right hat in the right town can build you an instant camaraderie with a customer. This builds trust and trust builds business.

That hat also will help you keep other things like glue and joint compound out of your hair while you are working under the house. At least keep a brush or comb in the truck to touch up between customers.

Leave any of your excessive piercings and jewelry at home. Cover up at least some of your tattoos if they will make you look weird to other people. No, I have nothing against them; I just know what works. You can express yourself later, while your cashing your checks. Right now you want to make money. Looking odd will not help and will certainly hurt because it will make some people uncomfortable. Your customers did not call for a rock star; they called for a plumber.

Attitude

Attitude isn't the only thing attitude is everything!
-Unknown

I know we have already hit this once before but we need to hit it again. This will make more money for you than anything you do, whether you are the owner of a one-truck outfit or a fleet. Practice saying over and over , out loud "I love my job I love my job". This can actually be a fun thing to lighten up the mood when things are not going well. When you like your job you will do it better and that's a fact.

Part of your attitude is how you feel about your customers. Customers are your family. They are the only reason you're there. They are your paycheck. Treat each one as if he or she were your closest relative. You would do anything to help your mother, right? If you treat your customers with the care and respect that they deserve (even when they don't) you will get many referrals and many invitations back to their home. They will be more likely to ask you to do other jobs or repairs for them while you are there.

> *"The customer (client) is the most important visitor on our premises.*
> *He/She is not dependent on us. We are dependent on him/her.*
> *He/She is not an interruption in our work. He/She is the purpose of it.*
> *He/She is not an outsider in our business. He/She is a part of it.*
> *We are not doing him/her a favour by serving him/her.*
> *He/She is doing us a favour by giving us an opportunity to do so."*
> - "Mahatma" M.K. Gandhi

If you are the owner of a fleet of trucks and plumbers, your employees' attitude is key to you making more money. Your employees need a winning, positive attitude.

Employees with bad attitudes are always a problem. They will steal from you every chance they get. They can steal from you by not working hard, by taking long breaks or by going to the bathroom too often. They will call in sick when they are not and generally be slackers.

Sometimes these are just problem employees who need to be let go. Other times it might be something you can fix. If an employee thinks that he or she is not being taken care of, they may develop a bad attitude quickly. There are many ways to avoid this. If all else fails, get rid of the employees who are dragging down the rest.

One bad employee can cost you productivity in the entire fleet and stir up all sorts of bad feelings about your company. It may be a personality conflict or a personal problem having nothing to do with you. It does not matter. Your job is to run a profitable business, not to provide a job for people who should not be working for you.

"Bill it is obvious that you are not happy here and that is not good for either of us. I have no choice but to invite you to seek employment elsewhere. I really do hope that you find what you are looking for. Friday will be your last day."

Cast out the scorner and contention shall go out; yea, strife and reproach shall cease.
-Proverbs 22:10

    Your employees do not need to be pampered but you would be surprised to see what will happen if you as their supervisor or boss gave them a little "good job" once and a while. When was the last time you went up to one of your plumbers and said something like "Hey Bill that was great job you did for the Thompsons. They spoke highly of you and I appreciate it." You know it would probably floor old Bill. It would make his day for sure and let him know that the work he does really is worth more than his pay.

    Try this one, "Bob I know that Mr. Jones is real pain sometimes. I hear you did a great job handling him when he wanted to pick apart the invoice. That's the kind of thing that keeps our customers calling us back. Good job."

    It makes people feel good to be appreciated. This helps you grow your business and keeps employees from jumping to another company for just a few cents per hour increase. Let your people know they are appreciated and they are more likely to stay with you.

    In the profit booster chapter we will discuss other motivators that will help you create a profit powerhouse without greatly increasing your overhead. Until we get to that point, smile.

    Some customers just like to argue. You know the ones. They have an opinion about everything and want to hear yours. At least until you give it. Then they want you to hear theirs and all of their reasons that you are wrong and they are right. Don't fall for it.

    Simply said, when the customer wants to argue spot him right away and do not be lead into it. If you know you're dealing with a person like this, control the outcome by letting him argue with himself. Arguing back will win you nothing, including the referral.

A brother offended is harder to be won than a strong city: and their contentions are like the bars of a castle.
-Proverbs 18:19

The Lady Plumbers

This is a good place to recognize that all good plumbers are not men anymore. All companies should be on the lookout for female plumbers with good work habits, just as they would the men. Any employer who gets a chance to hire female plumbers would do very well to do so. (Assuming they are good plumbers)

If any of you readers are of the fairer sex and considering getting into the trade, you certainly could do a lot worse. You will have certain advantages just being who you are and can expect a great career. Of course you might run into the usual chauvinist once and while, but most coworkers will be supportive.

The many benefits of having (or being) a female plumber should be obvious. Many of the service calls these days are placed and handled by wives. Some women are more likely to be comfortable with a woman plumber and in some cases the interaction may go more smoothly. Add to this a few good sales skills and you will have (or be) a highly profitable plumber.

Any male plumbers who work with female plumbers would do well to treat them with the same respect as any other co-workers. For our trade to continue advancing we need all the good plumbers we can get. We need them out there doing the job well and making our customers happy. There is certainly plenty of work to go around.

Chapter 3
The Truck

Set Up

Setting up your first plumbing truck is not hard. State auctions, which are held near most major cities, have many vehicles that have been at least moderately maintained. They are normally sold at below book value.

I recommend vans. The main reason is that you can carry many more things and keep them out of the weather. You will really appreciate this the first time you have to haul a new toilet or hot water heater around in the rain. A van also allows you to set up bins for organizing your equipment. It will then be readily accessible on the job. Remember this; **anything that can save you time makes you money**. Work vans are usually less expensive than a pickup truck or other type of vehicle.

Some plumbing companies prefer the box trucks or a step van. The step van is a good choice as you can carry more supplies, but the costs of the truck will probably be higher and the operational cost of gas and maintenance will also be higher.

The step van's low deck height makes it a good choice for getting in and out to get to your supplies; you can also easily outfit the step van to carry a lot of items.

For safety reasons, I hesitate to recommend the box truck. If you have to be jumping or climbing out of a box truck in order to get your supplies, the toll this will take on your knees will be high. Chances are you will have knee problems which will take you out of business in the future. Also, the time lost climbing in and out is not worth the trouble.

Another thing to keep in mind when choosing your vehicle is the areas where you will be driving. Many parts of older cities have narrow streets and driveways. If you have to park in the street because your van won't fit in the driveway, you will lose time at those stops.

You can normally find a work type van or step van that has been well maintained and is in decent shape for about one half of the book value. Anything less than book value of course is a good deal. Once you get the truck and have a chance to check it out you will need to start planning how you will set it up.

You should at least install a barrier or "headache rack" between the storage part of the van and the driving compartment. This will save your life in the event of an accident. It will also keep items from hitting you in the event of a hard stop.

The next thing to consider adding is some more lighting. If you plan to do any work in the dark those little dome lights just won't cut it for long. Adding a couple of low watt lights will keep you from fumbling for parts. Again, saving time makes you money.

A ladder rack or pipe rack is a must. This is the place that your long items or piping will be carried, as well as a place to hang other things for transport on a temporary basis. A rack can be a great place to haul large boxes of awkward but not too heavy materials.

Remember that PEX and CPVC should not be exposed for long periods of time to the sun's UV rays. These items should only be carried as needed on the pipe rack. UV rays will also attack PVC and ABS, but not as fast and without losing as much of their strength as the plastic pressure piping will.

Safety must be considered when you are using the roof rack. Anything that is put on it has to be tied down so that it cannot fly off in the event of a panic stop or accident.

Rubber tie down straps and even short pieces of electrical wiring will hold most plastics securely and will also hold some smaller amounts of steel piping, up to about 2". There are other strapping options, for example motorcycle straps and ratchet straps which may be handy for some loads, but the good old rubber strap is less likely to slip.

Never use rope to secure a load. Rope is much more likely to slip than straps that use the stored energy in the "stretch" to hold the load in place. If you have to haul more than a few pieces of steel pipe, have them delivered to your job in order to avoid an accident.

Piping that is not securely fastened can kill someone very quickly if it shoots off that rack. Steel piping can begin to slide easily. Once it starts it will not stop. The danger of it killing someone is very real.

Any very heavy loads should be delivered to your job. Adding several hundred pounds of steel pipe to the top of your van makes it very top heavy and can cause you to lose control of the vehicle. Not to mention that rubber straps can break and wire can slip on heavy loads.

Large diameter PVC pipe in big quantities presents another possible problem. The pipe lengths on the inside of your bundle can easily slip out under a hard stop. To be safe, have every pipe touching the rack or the strapping itself and not rely on friction from other pipes to hold it in a bundle. To avoid this, split the large bundle into smaller ones with their own straps. If in doubt, use extra straps or have your order delivered.

Most work vans have a long wall behind the driver and a set of side doors or a sliding door on the passenger side. To make the best use of your space, build or buy a set of bins for at least the long wall of your van. These bins should use all of the available wall space from the rear doors to the headache rack and from floor to roof. The pre-manufactured metal bins are the best. They are bit on the expensive side for someone just starting out. There are many different types to choose from. Choose one that best suits the type of work you are going to do. To save money, you can build the bins yourself. The few hours in the driveway working on it will save you several hundred dollars.

Having your materials readily available will help you make every visit more profitable. You will not lose time by routing around looking for something.

It is also a good idea to have one of the bottom sections devoted to frequently used power tools. This will keep them off of the floor where they are subject to damage or getting buried under other items where they are hard to find.

This is how to make a great 4' X 8' bin unit to fit into your van. As all work vans differ in shape from model to model or year to year, most of the measurements have been left to you to take from your vehicle. As long as you double check the uprights of your shelving unit as you go, there should be no problem fitting these instructions to your van.

Make a template for the ends and center of the bins by using a standard 1" X 10" X 4' shelf board and carefully measuring at different heights to draw the curve of the van's sidewall onto the board. Then use a jigsaw to cut the shape out. After trial fitting and trimming the first board to the side of the van, all the way down, use it as the template to create the other upright parts of your bin unit. Three uprights will do most of the time.

Once you have your uprights cut use 1" furring strips to set the shelf supports where you need them. The first shelf up from the bottom should just sit on top of or just clear the wheel well. This gives plenty of room to slip more boxes, bins or other items under the first shelf. Use drywall screws of the proper length for fasteners.

The second shelf should be about 12-18" above the first shelf. If located below the side of the van where the side narrows, it will be full width or close to it.

Once you have the uprights and the shelves in place, flip the unit onto its front and fasten a ¼" X 4' X 8" sheet of plywood to the back. This will keep your items from falling behind the unit and getting lost and will make the whole setup very sturdy. To make fastening the back of the bin unit go quickly you can snap a chalk line along the back where the shelves are to make sure you hit them with the screws.

Once the back is on, fasten 1" X 3" boards to the front of the shelves to keep things from sliding off. The entire unit can be slid into the van and secured into the corner with angle brackets and self drilling ½" screws available at the home centers. Make sure you have it where you want it before screwing it in. Only use the screws on areas where they will not go through the sides of the van. (Otherwise you might end up with a van that looks like a porcupine on wheels from the outside.) I like to leave just a bit of room at the rear of the bin unit for other items. Make sure you use enough brackets to secure a fully loaded bin unit. I would recommend at least 10 brackets and have them at every possible point to make sure your bin unit won't move.

Now that you have your new bin unit installed you can buy plastic bins for about $2 each at the home centers. They will fit in perfectly. Get the plastic bins as you need them, or load up now so you don't have to worry about it.

Think carefully about organizing your bins. The smaller more used items should be near the rear of the van so you do not have to climb in to get at them. Put the smallest items in the top bins. I like to have my frequently used power tools on the bottom shelf near the door. The bins above them are for faucet and toilet parts like washers, seals, flappers, fill valves, etc.

One of the things you use the most will be supply tubes. A great way to keep them handy is to take a piece of 1-1/2" PVC with a test plug glued into the bottom about 12" long. Cut the top at a taper and use one drywall screw to secure it to the rear upright of your shelf unit. Several of these on the rear upright will keep you from rummaging around for those supplies when you need them.

The space between the bin unit and the rear of the van is a good place to slip in your cordless drill case, water key and other items. A water key fastened with Velcro to the rear of the unit will come in handy also. More on that later.

When you are setting up your van, don't overlook those back and side doors. Most have a cavity in them, which is a great place to keep your rags, cleaning items, duct tape, gloves or other things. A spray can rack on one of them for holding paint, lubricant or other aerosol can type items is also very handy.

Keeping the inside of the van neat and orderly will add to your efficiency. It will save time on the job. The less cluttered the van is, the safer it will be for you to move around in it.

The Exterior

The outside appearance of your van is important. It does not have to be brand new. If you are doing a great job and just starting out, you can get away with not having the best looking truck in town. At least it should appear clean from the outside and should not have a lot of rust holes and dents or things falling off it. Remember, the better it looks the better first impression you will give. Once you get established, having a truck in good condition will help you when you are servicing the better neighborhoods.

If you are setting up a service truck it is always a good idea to have it professionally lettered with your company name and contact information on it, especially the phone number. People get used to seeing you, your van and your phone number in their neighborhood and new customers will call you.

Supplies

When you are running a service truck, you have to stock it so you are your own supply house. You never want to interrupt a job because you have to make a run to the supply house or home center. Getting your truck set up for this is not difficult but it will take some time. Give a little thought to the type of work you plan to do and the type of homes you will be servicing. It will go a long way.

If you are just starting out, you will need to come up with a list of supplies to keep on board at all times. As you go through your service calls you will need many items, which you will not have on board. If these are things you are likely to need again, buy several at a time. If you buy parts in quantity you will quickly build your stock to suit your needs. Your customers will appreciate the fact that you have what you need without costing them money while you run all over town shopping. Appreciative customers mean repeat business. You will get a reputation for always having just the right part on your truck.

The items you are likely to need repeatedly are: Fluid Master 400A fill valves, flappers, Mansfield and Fluid Master seal kits, wax rings, closet bolts, supply tubes for toilets. Include parts for older toilets that may be common to your area, like the old lift rod and tank ball styles.

It is a good idea to have a supply of chrome and/or plastic drain parts, faucet springs and seals, seats. Faucet stems for faucets common to your area are also good to have on board.

It is always a good idea to stock a couple of icemaker kits. Having a roll of ¼" PEX, a couple dishwasher elbows and dishwasher tailpieces for hooking up dishwashers will be good to have.

You need to carry at least one, preferably two 1/3HP garbage disposers. The ISE Badger 5 is great for this. It also helps if you are equipped with an appliance cord kit and a couple of electrical fittings to wire them with, along with a box or two of different size wire nuts.

If you have kitchen sink sprayers in white and black, you will be able to sell them on the job while you are on the job, as opposed to coming back. Keep at least one each on your truck. (Don't forget the Delta diverters too.)

At a minimum, select the types of raw materials to carry and use on your truck. PEX and CPVC are the most used products for water lines in many areas. Keeping a small supply of fittings and pipe on board will have you ready to make emergency repairs and even routine connections without special supply house or home center trips.

PVC piping and fittings are good to have on hand also. Having an assortment of Fernco couplings gives you the ability to make repairs using PVC pipe in most waste systems without having to stock a bunch of other stuff like ABS, galvanized piping or cast iron.

Of course no plumbing service truck is ever complete without a full stock of copper pipe and fittings in ½" and ¾". An assortment of these sizes of copper fitting does not take up much room and they really come in handy during a cold snap where freeze-ups are common and after hours calls the norm. Having your truck well stocked puts you in a position to capitalize on these times to your best ability. The goal is to get to a point very quickly where you have a complete and controlled inventory of everyday parts on your truck. This will make your work time more efficient and get you to the next job promptly.

Keep a list of everything you buy to put on your truck and mark it off or change your quantities as you sell the parts. This will help you control the inventory without a lot of extra work and will also help you avoid the embarrassment of running out of something while on the job.

While you are stocking your van, create a field price list of materials so when you are writing up your invoice, you have all of your pricing information handy and up to date. We will get into that in detail in our bookkeeping chapter. For now though, you need to have a working price list on board at all times so you will have consistent, accurate and profitable pricing without having to guess.

Restock your truck as needed. It will get to be a weekly event. Keep in mind that your goal is to go to the supply house when you have planned it, not because you ran out of a common item.

Microsoft Excel is a great way to produce these lists. You can set up separate sheets for toilet parts, sink and faucet parts, PVC, CPVC, ABS, Copper, PEX, trim or anything else you want. By having them on the computer they are easy to change as your costs change. This enables you to reprint the list to keep it current. If you do not know how to use the basics of this program have someone who does show you. You can enter formulas with your mark-ups so that all you have to do is enter your current cost and all of your unit tax and mark-up will be figured automatically to get your sell price for each item.

We will get into the basics of using Excel in the Records and Bookkeeping chapter just to help you get started, but if you really want to take advantage of the program, there are good books available. The program can do many great things for you, including help you estimate larger projects.

Chapter 4
The business Plan

Whoever fails to plan, plans to fail
-Unknown

The Plan

A business plan is essential. Whether you are just starting out or are running a multi-truck fleet, you have to have a plan. If you have a service fleet, meet with any and all lead personnel including your sales staff, field foreman, inside sales and logistic personnel and create or revise your plan. First we will go through putting together the plan for the one truck service operation and then we will get into how to prepare a plan for the multi-truck fleet.

Every business plan will be different, and most will be successful. The reason is simple. You know your area and customer type better than anyone else does. This is the type of knowledge that you use to formulate the plan itself. If you want to really see how successful you can be, just a take a little time and put a business plan together. Putting a plan together may sound like something only a college business major can do but you already know more than you think you do. Formulate your thoughts and write them down. Then you can evaluate them and get them organized.

Begin by securing all the necessary business licensing. List each one along with the cost of each. Check them off as you go. Don't forget to include your General Liability insurance on this part of your plan.

One of your priorities will be choosing the name of your new company. Choose the name to reflect the type of work you want to do most of the time. Are you going to concentrate on just plumbing, or will you be doing both plumbing and heating? Maybe you will be a full service plumbing, heating and air conditioning business.

After you have done these things, your next step is to plan the purchase and setting up of your vehicle. It's important to include every cost you can imagine. You don't want any surprises. Include your inventory planning and pricing in this part as well as all other costs. Don't forget insurance here either. Business use insurance is different from your personal insurance so be clear about your needs with your insurance provider.

The next step in your plan is marketing. This is where you will decide how you are going to go about finding your customers, what type of customers you want, and how you are going to convince them to do business with you. How can you attract the "right" kind of customers?

Take into consideration the type of work you will be engaged in, the area where you are going to be providing your service, and any other data you can come up with relevant to finding good repeat customers. Plan how you will roll out your business, slowly or all at once.

If you are a multi-truck fleet, hold a meeting with all of the lead personnel and go over all of the strengths and weaknesses of your field personnel. With this information you can play to those strengths, work on the weaknesses and build your business faster and better.

Tailor your plan to the people you have right now. If there are changes to be made you will recognize them and deal with them effectively. Remember those female plumbers too. You may find them better suited for certain calls, depending on your customer base.

Invoices

Of course you will need to include invoices in your plan. Think about how you will bill for your services. As simple as these little things may appear, your invoices are actually a source of profit. Choose an invoice or create one yourself that allows you the flexibility to break down your billing. Include separate columns or sections for "materials", "labor", "miscellaneous" or "other" as well as all the usual job information.

A good invoice to use and the one I used for years is the Adams NC2817 carbonless 2 part invoice. These are readily available at your local office supply store and will give you everything you need to bill your customers. You can even run them through your computer's printer to personalize them with your company information, logo, slogan or whatever you want. Your printer will only print on the top copy of course, but that is all you really need as the bottom is for your records anyway.

This type of invoice structure helps prevent you from missing anything and clarifies charges for the customer. It also helps to explain why the bottom line price on the job is what it is. Your copy of the job can be used to track profit and loss and anything else you need once you're back in the office.

It is necessary to have an invoice for every job. This is your protection against a callback after the normal one-year warranty of your work. It also gives the homeowner a piece of paper they can refer to when they need something else done and want to call you back as their plumber.

Whether you have your invoices printed professionally or do it yourself on your computer, put your complete contact information on it. Include your address, phone fax, and e-mail. Also include your logo and you slogan if you have one. This helps people remember you. Repeat business is the only way to make the big bucks in this field.

If possible, make sure your invoices match your business cards and your truck lettering. This gives a very professional appearance and makes a great impression.

Pricing Your Materials

"A cynic is a man who knows the price of everything, and the value of nothing."
- Oscar Wilde

The materials (inventory) that you will use are a big expense. Buying enough everyday repair parts to stock up the truck and to be prepared to do business is expensive. The mark up you place on your inventory is partially to cover all of the costs associated with having them on board. Carrying costs are included in this. It's difficult to accurately estimate carrying costs. You will spend hours each week buying, pricing, cataloging, shopping for and storing this inventory. The cost in terms of your time is worth quite a bit.

The smaller the item price the more mark up you need to calculate. For instance, if you buy a set of closet bolts for $1.75 at the supply house, you had to use your time to purchase them. (Let's say 30 minutes at $75/hr = $37.50) You paid for the gas in your truck to go get them. (2 Gallons of gas at $1.50/gallon=$3) Then you have to spend time to price them (15 more minutes total = $15) and then you will tote them around on your truck for an unknown amount of time before you sell them and get your money back. (We'll let that one slide because we are such nice plumbers.) Still the total value of those bolts is about $55! There are other costs associated with that part too. Soon the invoice will arrive. Someone has to handle it and pay it. If the invoice was incorrect, someone may even have to spend time on the phone discussing it. All these costs are associated with that purchase.

Now, you're sure you can't get that much for them but all of a sudden a $5.50 price tag does not sound too unreasonable. Of course you will be buying them by the dozen to keep your costs down and your profits up. You will rarely buy just one thing on a trip to a supply house.

All of your low cost materials (under $8-$10) should be priced this way, at three or four times their costs. Otherwise you will be losing money on many of them. There is no reason to sell items off your truck or from anywhere else at less than a 25% mark-up. If you do, by the time you figure all of your actual costs, you are just passing that item through your books. If you cannot get at least 25% on something, you should not supply the item. In many areas of the country, due to higher tax rates or other factors, 35% is the minimum markup on supplies and equipment provided and installed by plumbing contractors. Let your customer bear the total cost of every purchase.

Develop a guideline to calculate pricing on all your items. I would recommend the plumber just starting out in business to use something like this:

Item Cost	Mark up
$10-$15	100%
$16-$25	75%
$26-$50	50%
$51-$100	35%
$101 and up	25% Minimum

Your inventory will include many items which you will sell off your truck and which will cost much less than the minimum $10 shown on this table. You will need to determine the cost of those items for yourself. For example, you buy washer assortments, which have large quantities of washers with an actual unit cost of only a few cents. You should get at least $1 each for these or more. Similar items, like closet bolts, are about $1.75 at cost. You should sell them off of your truck for around $5. A $.39 wax ring sells for about $4.50. A Fluid Master 400A that you buy on sale for $4.50, you should sell somewhere between $12 - $18. These are general guidelines. Your market may have different factors to consider and you will charge what the market will bear.

Eventually you will have to deal with a customer who starts picking apart your material list on the invoice. Ever hear this one? Customer: "I could buy that at the home center for less money." There is no need to be rude when responding but the customer should have it pointed out to him that you have his best interests at heart when you stock your van for speedy service ahead of time.

Here is a good response. "You could, but I did. This is my price and it covers all of my costs. I spend hours every week going through my stock and going to supply houses on my time to make sure that you do not have to pay my labor rate while I run all over town looking for the lowest price on an item that only costs a few dollars. Or worse yet, pay to have me wait while you do it. By having these things on my truck at this price, I cover all my carrying costs and save you hundreds in wasted labor fees. Also, when I provide it, I have to warranty it. If something is wrong with anything that comes off my truck, I can get another one and the problem of the bad part is mine."

This usually will satisfy these customers. Explaining it to them is always a better solution then getting offended and arguing about it when they question you.

There are other items that you will use regularly that are not used up on the job. These items are referred to as "expendables" or "ancillary" items. Examples of these items are solder, paste sand cloth, acid brushes, acetylene gas, mapp gas or propane tanks, rags, cutting oil, pipe joint compound, plumbers putty, cleaners or other similar items. Your customer should be charged for these things. Not only is this a profit booster, but in order to do the work you need to make these purchases.

A good rule of thumb for a service job is $1 for each type of item used on the call and a $2 minimum whenever these products were used. I list them as a line item on the invoice as "Expendables". If your invoices have a "Miscellaneous" box, this is where it should go. You can write out each one of them if you would like to avoid the customer asking you about it. (It works!)

Here is an example of how to bill for "Expendables": Let's say I put in a new kitchen faucet that required soldering on a new valve under the sink. I also had pipe dope used on the joints and plumbers putty on the faucet flange. On that invoice, I'll have Expendables: solder, paste and sandcloth, putty, pipe joint compound and rags. In the "Miscellaneous" box I write "Expendables" and include these items at $6. This guarantees that I will be paid for all of these items. These items are not free. It is appropriate that the plumber be repaid at a profit.

If it is a larger job requiring more of these things than a simple home repair, I figure 20% of the raw material cost. The best rule of thumb for the plumber in the field is to look at each system of billing for these items and to use either the $2 minimum, $1 per item, or the 20% rule, whichever is the most profitable.

You might think that this is being a bit nit-picky but think of it like this. Let's say you average a charge of about $7 per job for expendables all year and you do about 800 small jobs a year. (Probably on the light side for many service plumbers) That works out to $5,600 that you would leave behind! Not too "nit-picky" after all is it?

Setting your Rates

The last part of your planning stage is to set your labor rate. This is also a good time to decide whether or not you will charge a "Service call" fee (a minimum rate to show up). Some call this a trip charge but it means the same thing.

Let's look at the hourly rate first. Find out what the average rate in your area is and start there. Discounting your services will only attract more of the "penny pincher" crowd. You need customers who are willing to pay full price for good service. Selling yourself short is counterproductive at every level, especially if you are starting out. Don't do it.

We all know the customers who think you are stealing from them when they get your hourly rate. You know the one who says "I worked hard all my life and I never made $65 per hour!" Guess what? These people will say the same thing to you if you are $45/hr, $55/hr or even $75 or a $105/hr. These people will grumble anyway so all you can do is smile and nod and hope the check doesn't bounce. If they push you can invite them out to your truck for a look at why you are paid that rate, but it is best not to argue with a people like that. You have other important things to do and other people to serve.

Call several of the better-known plumbing operations and get a labor rate for each one. Choose the average one. After all, you are worth as much as any of them aren't you?

If you are considering starting out with lower prices consider this. Will your doctor see you for less money because the doctor down the street will? How about your dentist? Maybe the guy that fixes your car would cut his rate. You know, I'll bet your lawyer and accountant would love to lower their prices as well since you were nice enough to charge less.

Of course none of these other professionals would even think of cutting their rates. Neither should you. If you cannot get paid what you are worth, go fishing or play golf instead.

In fact Donald Trump renovated a posh apartment building only to find that he could not get tenants to move in. His rates were competitive with the other luxury buildings in the area. His solution was to raise the rates to well above the others. Very soon the entire building was sold out!

Even in plumbing some customers will choose to do business with the most expensive company in town due to the perception (and usually the reality) that you get what you pay for.

With the labor rate established we have to decide what the minimum charge will be. That is where the "Service Call" can come in.

As soon as you schedule an appointment, from the minute you leave your last stop you are working for the next job. You are entitled to be paid for the time it takes to get to that job. There are several ways to set this up.

The first (and least popular with customers) is the "Service Call" This is a rate that is charged automatically when the appointment is made. The regular time starts once the plumber is at the job and begins to work. The purpose of the "Service Call" rate is to cover the drive times between jobs. This is fine and it is fair if you know your average time between stops but customers have a hard time with this.

A better rate that the customers understand is having your hourly rate be the minimum charge. Start your time on each job as you leave your last stop. This ensures that you will be paid for all of your time and that you will make enough on your shorter repairs to be worth the trip.

Some companies are moving toward the "Flat Rate" structure of doing jobs. They work by using known averages of times and materials needed to do a given job. By doing this they are weighted with a bit of a buffer to cover those times when things do not go as planned.

These rates are available as software packages or as books. They tend to be more profitable to the contractor than other rate structures. You will find that these packages of information can be a bit costly but if you are planning to structure your business this way, they will be well worth the investment.

Customers think they are better protected because you have "set" rates. Sometimes they are of course because we all know that not everything goes as planned. The majority of them, however, will be paying more for a job under this structure. The whole purpose of having this type of price structure is to protect the profits of the contractor, with the secondary benefit of protecting a relatively few customers against fear of the unknown.

Many customers prefer this method because they feel less likely to be taken advantage of by the plumber doing the repairs. Customers tend to gravitate toward things that make them feel comfortable. For that reason I believe anyone who uses this type of billing structure will do just fine. Not to mention that most companies that move to this structure report a measurable increase in profits.

Once you have established your rates, immediately establish a policy to explain your rates to every customer when you set up the appointment. Be very sure they understand and have the chance to ask you any questions before you get all the way to their home.

Sometimes people will ask your rate when they call. You may have to broach the subject yourself, but do it before they hang up. Do it before you actually set up the appointment. Something like this;

"Next Tuesday? Let me see. While I get out my appointment book let me explain my rate to you. It is $75 per hour plus any parts I have to supply from my truck to make the repairs. There is a one-hour minimum charge, but no 'Service Call' charges. My time starts when I leave my last location. Do you have any questions?"

At this point they may ask about how long you expect their repair to take. Give them as honest an answer as you can.

You; "Barring any unforeseen circumstances, I can most likely have your toilet good as new in about 30 minutes or so. I would not expect it to run past that first hour."

They may ask you "What 'unforeseen circumstances'?" You will respond:

"Well Ma'am I have never been to your home before. I have no idea what condition your plumbing is in or what other things may be wrong with your toilet that could be part of the trouble. Most of the time plumbing is in fairly good condition, but there's no way for me to tell from here."

Now both you and your customer know what to expect. There has been no double talk. You've been completely honest. You have protected yourself and when it's time to deliver the invoice it goes much more pleasantly.

Sometimes you will get a customer who calls and asks you your rates only to announce to you that ABC Plumbing down the street has a lower rate. Something like "ABC Plumbing only charges $55 per hour!"

Never be rude to these customers even though you know what you are dealing with. Handle them by saying something like:

"Wow, that's a great rate, I hope they do a good job for you if you choose them. Our rate is "X" if you would like to make an appointment I can get you on the schedule for next Tuesday."

They may even want to challenge you by asking, "Why are you so much more money than they are?" The only type of answer that will work (without being rude of course) is:

You; "I don't know what their costs are or how they base their rates. I can only be responsible for my own rates and they are firm, set rates. If you prefer to deal with us we will be glad to help you."

Don't forget to thank them for calling, either way. If the other company botches the job, they might just call you back. Don't be baited into arguing with these people. They will only waste your time and if you do get a job from them before it's over you might wish you had not.

Special Rates

What about nights and weekends? While you are just starting out, there is no need to charge extra for these calls if it makes you uncomfortable. Not doing so will bring you more new customers but once your business is established, go to a rate structure that allows you to get paid "overtime" for your work during these odd business hours.

This will serve two purposes. One is that it will encourage people without emergencies to schedule you during regular hours which, of course, is more convenient for you. The other is that when you do work during these times you will be paid a premium for your trouble.

You can be as creative as you like on structuring your rates. Keeping things simple is the best way to go, but a little creativity may give you a better chance of making more money.

For instance, suppose you set it up so that any work done after 5pm on weekdays is at a time and a half rate and any weekend or holiday work is at double time. You will be surprised at the people who will pay your time and a half rate for their convenience. The weekend and holiday double time rate will keep you from working to death but the times that you do you will make enough money to make it worth it!

Tool and Equipment Rates

Many of the large pieces of plumbing equipment you will use come at considerable cost. These items are part of every day business and can be considered as part of your overhead. Many types of equipment have limited use before they are going to need some kind of work, repair or maintenance. For this reason, many companies have tool and equipment use rates that are charged as a miscellaneous cost on the bill.

Some companies charge for the use of things like ladders, power tools, drain machines, lights, generators, and everything else that is used on the job and is not a basic hand tool.

This may sound a bit extreme but it really does work to protect your profit. If you spend $350 on a drain machine and use it three times before something happens to the cable and then have to pay another $125 to repair it, how much money can you expect to make by owning that piece of equipment and not charging for its use? There must be a charge for the use of this type of equipment or it is actually costing you money to own it.

Of course some power tools are more durable than others and less prone to breakage, for example, a circular saw, drill or sawz-all. For these pieces of equipment list the blades or bits under expendables and put a $2/each charge on them to cover those costs. At these rates you are not getting rich, but you are covering your cost so that you can keep your profit. When you do need to replace these items you are not going to lose money. If you choose to charge a fee for these machines just keep the rates fair. It is not unfair to charge for them, but it is unfair to try to pay for them on every job!

Drain machines, mud pumps, jackhammers, threading equipment and other heavy-duty items are much more likely to need repair. Be sure you get a fair use fee for this equipment.

For a good idea of what that fair use fee would be, call your local rental center and get their rental rates for these items. Your charge should be close to or the same as the rental place.

Having these specialty tools on board and not making a profit from their use is not good business. Again, the whole idea of charging for the use of these items is to protect your profits and hopefully to increase them.

A brief word about drain machines. No homeowner should ever own one. To folks who are not trained and experienced they are quite dangerous to work with. I happen to also believe that most plumbers would do well not to own one.

Unless you're in the drain cleaning business specifically or have a crew on a truck that does this work regularly, a drain machine often is more of a liability to you than just forming a relationship with a reputable drain cleaning company to whom you can refer that work. Repairs on drain machines may cost you more than the money made when you do use one.

It's a different story if you're going to commit a crew and truck to that work. If drain cleaning is going to be a major part or focus of your business you'll do fine. If you are going to have a small business with only a truck or two going, let someone else do the major drain cleaning work.

If you choose to put a drain machine on your truck, purchase the highest quality machine you can afford. High quality drain machines are less likely to need repair, will be easier to maintain and be more likely to work well after sitting unused for long periods of time in your truck.

Do NOT buy a cheap, home center type drain machine. You are truly wasting your money and in the long run you will lose money by owning it.

Wipe down every tool you use on a job, power or otherwise, as it goes back into the truck. This keeps the build-up of oil, pipe dope and other residues from gumming up your tools. This alone will save you repair bills in the long run. If you have drain machines wipe down that cable. A regular cleaning and oiling is a must. Yes it is an unpleasant job, but just think how much you had to pay for it and the job will get a lot easier.

A can or two of carburetor cleaner and degreaser make clean up jobs go faster. You will of course be charging for these anyway under "expendables" when you need to use them.

Well-maintained tools will last longer than abused ones. With a good maintenance program you will be able to use these usage fees to increase your profits.

While we are on the subject of your tools a word or two on hand tools is required. Buy only high quality tools. Do not waste money on generic brand tools that most often do not perform well and can be dangerous.

Cheap chanelock pliers can slip and do damage to things being worked on as well as your knuckles. Pipe wrenches that slip are even more dangerous. Cheap screwdrivers break and ruin screws. These are just two examples of cheap tools which are a waste of money. There are many others.

Investing in quality tools means buying "Chanelock" chanelock pliers, Rigid pipe wrenches and Rigid basin wrenches only! Buy "Crescent" or Craftsman adjustable wrenches and Craftsman, Snap-On or Mac quality hand tools for things like combination wrenches, screwdrivers and other hand tools.

These tools are expensive. If you don't have them, accumulating them may take some time. You will end up spending thousands of dollars for proper tools to do the job. Treat these tools like the investment they are. Your tools, like your van and your appearance, represent you. Take very good care of all these important assets.

Quoting Larger Projects

You will most likely be giving quotes from time to time (or regularly) for larger jobs than your normal repair or replacement projects. Do not waste your time giving free estimates on small projects. You will almost never get them anyway. If it's a normal repair or light remodel, keeping it based on time and material (T&M) will be best for you and your customer.

If you get into larger or more involved renovations or new projects there are certain things to keep in mind while preparing your quotes. Get used to making a "Summary Sheet" which is simply a line-by-line cost description of everything associated with that project. Once you begin bidding work, you will develop your own summary sheet template, which will help you remember the important things that go into every quote. This is another great reason to get used to using Microsoft Excel, which we will get into later.

Include permit fees and any other fees that are associated with the job, including your time to file for the permits. Add your state sales tax on all items where it's required. Make taxes and fees the first few line items on your sheet so you will remember them.

It's important that you personally look at the job before giving an estimate. Be well aware of what materials and time will be required. If there is any doubt about how long it will take to do any part of the job, make sure you allow enough time in the quote to cover yourself. Once you have presented your estimate, you will not be able to recoup any expenses for work or materials if they were not listed in the original quote. Keep this in mind, you can always reduce the cost, but rarely can you raise it after the agreement.

Your quote should also include some kind of expiration date for its validity. The possibility of price increases needs to be taken into consideration. Thirty days is normal, 60 days will most likely cover you if you need to extend it.

It's important that you and your customer know fully what you will and will not be doing under the quote. Always include a "Scope of Work" section in your proposal where you will spell out all the details of the job.

Here is a sample of a summary sheet and a quote to give you an idea of what they may look like. This is for a light renovation consisting of moving piping for a new location of a kitchen sink, demolition and replacement of sinks and faucets in two baths and a kitchen. You may not even quote a job of this size, but it is a good example. You can make yours any way you want to, as long as it meets your needs. Your quote will be on your own letterhead, of course.

M. Scott Gregg

Project Summary Sheet	
Job Name: Jones Kitchen and Bath renovation	
Date: 3/15/2004	
Item	**Cost**
Permit	85.00
Sales tax	239.96
Fixtures	4,765.28
Demolition	1,000.00
Disposal of removed items	200.00
Pipe and Fittings	567.23
Expendables @ 20% of P & F	113.45
Total	6,970.92
Total Including Profit at 30% Margin	9,958.45
Labor 25 hours at $80/hr	2,000.00
Total cost	11,958.45

Quotation

Date: 3/15/2004
Customer: Jones
Address: 1501-Anywhere Dr.
            Anytown USA
Project: Kitchen and Bath Renovation

Scope of work:

Remove and dispose of all existing plumbing connected to the existing kitchen sink, including disposer and sink. Remove and dispose of all existing faucets and sinks in two bathrooms upstairs (3 total) and all necessary connected items. Relocate piping as required for new sink locations (2). After installation of new stone tops (by others) supply and install selected new plumbing fixtures, disposer, faucets and trim in kitchen and two bathrooms upstairs.

Total quotation amount including labor and tax: $11,958.45

Notes:

This quotation is good for 30 days.
Cabinets and stone tops not included.
Demolition of existing cabinets and tops not included

Quoted by:

Joe D. Plumber

Negotiation

There are many good books on negotiation available. Try to find time to read a few of them. Teaching negotiating skills is complicated, difficult and involved and would require at least another whole book. In the space we have, however, we can discuss a couple of negotiating points which are specific to our trade.

Sometimes the cost of a job becomes an issue with the customer and he/she wants to negotiate the price. This is a fine and perfectly acceptable part of doing business. If you develop good negotiation skills early you will enjoy working with your customers and will be able to come to an agreement that meets their needs at a price they can afford.

A negotiation where parties get together and develop a plan that fits a budget is always beneficial to you as well as to your customer. You may make even more money on the job (percentage wise) depending on what is eliminated. In addition, your customer is going to get a job that he can afford and he will know exactly what he is getting for his money. The customer may also choose to have you back at a later date to add to the project.

This is one of the most important paragraphs in this book and will be the one that protects your profits time and time again. Notice we have said nothing about lowering the price. Never, never, never lower the price for your job without taking something away from it! Now, read that sentence again. You need to remember it.

Did you read that again? Good. Cutting into your profits for no reason is extremely bad business and will cause the customer to think that you were overpriced in the first place.

Look at it this way, if you go to the grocery store and get $110 worth of stuff and only have $100, you have to put something back to lower the bill. That's just the way it is. That's also the way it is for any successful business.

Involving your customer in negotiation helps keep him or her from calling your competitor to shop your price. Everyone knows that if you are the last one to bid a job and you know the other prices, you can get that job if you want it. Your best bet is to negotiate that job onto your books and off of the street as quickly as possible to avoid losing it.

It may go something like this:

Homeowner: "Gee, I really didn't think the job would cost that much. I really can't afford that. Is there any way you can do the job for less money?"
Plumber: "I don't know, let's go through the job and see if there is something that we can do differently. Maybe we can use a different faucet or leave some of the work for a later date. How much do we need to cut out of it?"

Now you have gotten your customer involved in the process and you will be able to work with him to get a job he wants for a price he can afford. The best part is you haven't lost the job to someone else. The customer doesn't need to see your costs or summary sheet but you can go through those items with him anyway. Maybe the customer will handle the disposal of removed items, or switch faucets. Either way, both of you can benefit from the negotiating process.

The other advantage to this type of negotiation is it adds to your professionalism and to the image of our trade. We are not crooks just because we make good money. We make our money honestly by providing good service at a competitive price. We do not cheapen ourselves by lowering our rates.

There are times when you may need the work more than others, or need the work to keep a couple of good men busy. That's fine, but consider that when you prepare the quote. Don't let yourself get into a position of cutting that price without taking something out of the job in return.

Any good negotiation results in two (or more) satisfied parties. If you enter it trying to take advantage of someone, most of the time the results will not be good and it is just not necessary in our trade. There are plenty of honest ways to make money.

Chapter 5
The Plumbers Network of Customers

It's a poor dog that won't wag his own tail
-Unknown

Marketing 101 (For Plumbers)

Wether you are a one-truck start-up or a 100-truck fleet you need to market your business in order to find and retain the kind of customers you want. Your marketing should work with your company not against it. Word of mouth is a great tool; the referral is something we strive for. If, however, you are serious about being in this business for yourself or to develop a company you need a marketing strategy that will work successfully for you to get that phone ringing.

Business cards, stationery, truck lettering and invoices are all basic tools for marketing. These items should all match to give a professional appearance to your operation. Give a business card to every person you meet who, through conversation, finds out you are a plumber.

Always have some of your business cards with you. Keep them in the little holder that comes with the box of cards to keep them fresh. A shabby business card is not acceptable and will be thrown away a lot faster then one in good shape. This is part of your first impression so always make it count.

One of your better marketing tools is the decal with your company information on it. You can get them made by the thousand-count roll for surprisingly little money. Put them in an inconspicuous location on every piece of equipment you install. Put them on the front of the garbage disposers, on the water heaters, on the bottom of sinks and anyplace else where the homeowners will see them when they need you again.

The types of marketing you do will have a lot to do with the type of customer and work you will get. The best type of customer is the one who has a high-income level and does not have a lot of time to solve his or her plumbing problems. They just want you to get in, fix their problem, or install their new item and go away.

These are the working professionals like the doctors, lawyers, real estate professionals, brokers, etc. They want to be treated fairly and with respect and do not appreciate a plumber who walks into their luxurious home and sees a blank check. Trying to trick these people out of their money will get you a bad reputation fast if not thrown in jail, but developing a friendly and trustworthy reputation with them quickly will get you many of the high quality referrals that you need to make more money in this business.

Finding your customers is the easy part. Keeping them and getting that referral is the part that separates the highly paid, successful plumber from everyone else. If you are following the advice on how you present yourself and represent your company these things will fall into place nicely.

Define Your Customer Base

The first thing to consider when deciding on your marketing plan is what type of customer you want. Once you know that, your marketing effort becomes a lot easier.

Do you want to get calls from anyone and everyone who has a plumbing problem? If so then the first thing to spend your marketing money on is a Yellow Page ad. The bigger ads or ads with color will get you more calls, but will cost you more money. Having this ad in place will get you a lot of calls, especially during a particularly cold spell when many pipes freeze and everyone books up quickly.

Explain your available service in the ad to generate those calls. Words like "24 Hour Service Available!" or "Night and Weekend Rates!" are very effective. There are people out there who do not want to miss work and will pay a higher rate for those hours even when they do not have an emergency. You want those people to call you before they call any other company. All they need is a reason to do so.

If you want specialized customers then the Yellow Page ad may be counter-productive. The same yellow page ad which gets you all those calls will also get you calls from people who are price shopping and have little intention of using you in the first place. You will get calls from every area covered by that Yellow Page book. That may mean you have to drive all over town for a few calls.

If you are a fleet plumbing company, you need and want this type of exposure. If you are a small one or two truck operation it is probably a better idea to focus your marketing in other directions.

Direct Mail

Direct mail is a great way to get your name to a targeted group of people. There are many direct mail companies that put together mailer packages for small businesses. These mailer packages break down the demographics in order to focus on the people you want as customers.

There are two types of these programs. The first one will let you direct your mailings to a specific geographic area or part of town. This is handy if you want to work close to your shop most of the time, or if you know the areas where you will find the type of customer you want. You can choose mailing campaigns to work in a certain part of town, or to focus on a certain income level group.

There are programs that will send your mailer only to those people in the top 10% or so income bracket in a given area. This is particularly good for selecting people who like high priced items and live in luxurious homes. You can make more money installing one $900 faucet than you can on installing ten $40 faucets. These customers are your own private gold mine, but there are other ways of getting to them. We'll discuss that next.

Realtor Gold Mine

Lets say that you want to focus your efforts on an area of town where the income levels are relatively high and the homeowners are too busy to shop around for their plumbing needs. The best way to get into this area is visit the realtors who serve that area.

Find out which ones are the better-known real estate companies for that area and be prepared. Make up some fliers with your company information on them. Include the type of work you do, your regular and special rates, and ask to meet with the person in charge of each office. Start by asking them if they work with any plumbing contractor currently and if so are they happy with the service their customers are getting. If the manager is receptive, make an appointment. Do not be late. If the manager is not very receptive, get the address and mail him/her a nice letter along with your flier and a few cards. Follow up by phone in a couple of days.

Once you get the appointment, give the manager your flier and your card. Tell her your business is based on great professional service, reasonable rates and fast response. Tell this person you would like to work with her to solve plumbing problems for her clients. Let her know you have a fully stocked service van. Assure her that you usually have the items needed to make repairs in one service call and that you avoid costly trips to the supply house. Ask what the best way would be to get your contact information to her sales people.

Sometimes the realtor will let you do a short presentation to the staff before or after a sales meeting. This lets you distribute your flier and cards to a lot of key people at once and will result in many jobs almost immediately. It also lets you meet these people face to face so they can see what kind of person you are before they start recommending you to their valued customers.

This is a great example of where that first impression counts. If you are polite, well spoken and neatly dressed you will probably get many referrals.

Many times during a home inspection a plumbing problem will arise. The homeowner is often responsible for resolving these problems. Most of the time the homeowner will ask the realtor if she knows a good plumber. This is where you will get a job. If that realtor is comfortable with you she will refer you every time.

This very scenario is played out in your town hundreds of times every day. Only the plumbers who are proactive in getting to these people will get these calls.

Before you know it the majority of the calls you get will be from people who are selling their homes who need a basic (high profit, short visit) repair done and will pay you with a check that will not bounce. These are more profitable repairs than the ones that take you much longer.

Realtors also control other properties. They handle rental homes, apartments and condo complexes as well as office space. You will find you will even be getting calls and making repairs for the real estate company at their sales offices.

In the profit booster chapter we will discuss items that you should have on your truck to serve these special needs, as well and how you will go about selling these things to make more profit.

Local News Papers

The local newspaper is a great way to reach your prospective customers. Many of the larger metropolitan papers will only let you do certain things, but the smaller local papers can give you a wide range of ways to reach your customers.

Some of the smaller local papers have a place to submit your press release announcing you're in business. This is a good start, as people prefer to do business with local companies.

You'll get some business by placing small ads, but the best way to use these media outlets is to prepare and submit articles for them. Most of the time you will not get paid to submit these articles, but they are free advertising and they allow you to set yourself up as the " expert" on whatever subject you submit. It also gets people used to seeing your name associated with solving plumbing problems. Running your ad at the same time your article is printed is a great way to get calls.

The best topics for articles are ones that address problems you know are common to the industry. Low-flow toilets are still considered to be a big issue with people who have one of the first generation 1.6 gallon per flush models in their home. If you prepare a short article on these fixtures and detail how the new models work so much better than the older ones you will be replacing cheap builder grade toilets with higher end new models that work for sometime to come. Your customers will think you are a hero and you'll get a lot of referrals from them.

Some other great subjects for these articles include; hot water heaters, water hammer, hot water recirculating systems, home water treatment systems, garbage disposers, tankless water heaters and any other thing you can think of that will show you as the expert and get your name in the paper.

The biggest advantage you will have by writing these little articles is your customers will think of you as a plumber who is articulate and knowledgeable. Every opportunity you have to get your name in front of prospective customers is one step closer to making them YOUR customers.

Gifts/Handouts for your customers

Everyone has seen the refrigerator magnets many companies give their customers. What you might not know is just how good they work at getting your name out! Be as creative as possible with these things. You need your company to stand apart from the rest. There are many ways for you to accomplish this without going broke in the process. Use the shape of a work van or even a toilet. People will put these on the refrigerator next to the pizza delivery magnet and they will remember to call you again. One glance at the refrigerator and you are being called for another job.

Avoid wasting your money on trinkets, such as key rings, which will get thrown into a junk drawer never to be seen again. Give items that will either be used regularly, or greatly appreciated.

Other good handouts are can huggies with your company name, phone number and slogan on them. Miniature models of your truck are great but a bit pricey for everyday handouts. Even a T-shirt with your company information on it, or including a witty saying can be a great handout. Novelty T-shirts can cost less than hats in quantity. Think of someone wearing your shirt to the beach as a cover-up. Imagine a long shirt with a tool belt printed all the way around and pants riding a bit too low in the back. It could have your company logo front back or both. People would wear it for the novelty and you get your money's worth for the advertising!

A great example of a handout that is not only appreciated but will get you remembered and talked about is what a local company here in the central Virginia area does.

Kelleher is a large service company that handles oil, plumbing and heating and has become quite well known, growing from just a couple trucks to about a dozen or so on the road.

When Kelleher goes on a service call and encounters a child, the service person goes to the truck and presents the kid with a nice teddy bear, complete with a collar and the company information on it.

The trucks carry several of the bears on them. When the child receives it there is a huge impact! Nothing makes moms more pleased than to see someone sharing with their kid. The kid is instantly happy, no matter what else is going on in that house at that time.

The company has become quite well known for this gesture. Even though it costs them a considerable amount of money to maintain the program they have found it's just too good all around to give it up. Their vans even sport an image now of the little Kelleher bear on the back. Along with their great service, this has made them very well known. They've have built a very loyal customer base.

Referrals

So you found a great customer and did a super job on his project. He's so happy with the work he tells you he can't thank you enough. Yes he can! Ask for a referral.

Always, before you leave a home, give the customers some of your cards and let them know you would appreciate it if they told their friends about you. People like to help others, especially when they're happy with the service so why not just ask for the referral? You'll get a lot more business this way. Make getting the referral a goal on every job you do.

Solid Surface is Solid Gold

There are many companies in your area right now selling and installing solid surface counter tops. These tops include Corian®, stone, marble and other exotic tops for the homeowner who wants the very best. These tops can cost into the tens of thousands of dollars for one kitchen or bath!

Almost every time a top like this is installed the homeowner is doing a complete remodel, which requires new faucet fixtures and trim items. You can make a lot of money by supplying these items or at least acting as the middleman between the supply house and the homeowner.

Many of the installers of these tops do not do plumbing work so the homeowner has to choose his own plumber to do the disconnect work and to come back after the top is set and install all the plumbing.

The trim items that are chosen to go with these exotic tops are always going to be high end, high profit items. By supplying them to the homeowners you are able to provide warranty support for them and get your mark up at the same time.

Many times you will be able to purchase these items and re-sell them to the homeowner for less money than if he/she purchased them directly from a wholesaler's showroom at full list price. You are able to save the customer the trouble of dealing with the supply house. The markup will cover the cost of handling any warranty or delivery issues that may arise.

Find out if the counter top company is sending their customers to a supply house to pick out faucets. If they are, build a relationship with the showroom people. If they are not, have them send the customers to you. You may need to offer the solid surface company a commission for the referral. The terms of the sales can be negotiable. It could be a flat rate fee or maybe 10% of all sold items will make everybody happy.

There are two ways to break into this very lucrative business. You can either align yourself with the solid surface installer as a sub-contractor, or you can offer your services to him on a referral basis.

Begin by meeting with the owner of the company to find out what their current arrangement is. If they are already working with a plumber, find out if it's working out well.

Ask the ones who are not already using someone if they would be willing to use you as a referral. Many will like this relationship because it solves one of their problems. Hand them a generous stack of your cards and go to work.

The very best relationship you can have will be with the installer referring you on every job from start to finish. You pay them an 8

- 10% commission for all of your work. This gives you complete control of everything and you don't have to wait for them to pay you! You get loyalty and steady high-end business. They get to make a little extra profit from your work with no risk.

Consider taking the installer to lunch once in awhile to get better acquainted and build a good strong relationship. The better your relationship with the people who refer these customers to you, the better off you will be and the less likely they will be to allow someone else to move in on your territory.

Once you are doing the work associated with these counter tops you will be in some of the best homes in the best areas of your town. You will be working with people who are not afraid to spend money on better items or service. All you have to do is let them know of your willingness to help them out. This is where the "up-sell" comes in as described in the next chapter.

Your biggest concern when dealing with these customers is to be the best plumber they have ever met. Make sure you leave a lasting impression of your attitude, manners, and sincere interest and concern about their needs,

Do not avoid any opportunity to talk with the customer during the job. If you get one that wants to talk to you the whole time you are working, GREAT! By all means let them talk. Keep working though and do not let them slow you down. You have other things to do, take the time to listen and keep the conversation going.

If you are asked to do some other non-plumbing task at the home and you can handle it, do it. Help in any way you can. Find out if there are other projects coming up which will involve plumbing. Make sure that you give them a few of your cards.

Go the extra mile to ensure that your work is perfect in these homes. The drains are all perfectly straight, the travel of the faucet swings are dead even. The soap dispenser and the disposer are exactly where the customer wants them. Before you leave the job ask the customer to look at everything and make sure it's perfect. If all of this is done well, the customer will be happy to refer you to friends.

If you haven't done so already, begin developing a portfolio of photographs of before and after pictures of jobs you have done. You will not be able to get these pictures on every job but it will not take long to amass a very impressive portfolio.

Interior Decorators

This is perhaps the most overlooked source of good high quality customers for the plumbing contractor. Interior decorators deal with people who hire professionals in order to help them plan a renovation or a new home. The plumbing fixtures are as important to them as the drapes, wallpaper, furniture or paint. Using a plumber's knowledge of modern equipment and methods is a real advantage to a decorator.

Approach these customers just as you do the solid surface countertop people. The main difference is that these people will not want to use you as a sub. Their main interest will be someone who does a great job and will be a good referral for their customers. A portfolio will be very helpful in situations like this. Once the decorator sees that you do great work and can handle high-end fixtures worth hundreds (or thousands) of dollars, she will be comfortable referring you to her customers.

The Wholesaler's Showroom

The plumbing wholesaler's showroom is another good place to find great customers. People who go to these showrooms are looking for the best in high-end fixtures and appliances. They are not happy with the quality, service or knowledge at the local home centers. They're willing to pay for the best products and to deal with high quality, knowledgeable sales people.

Get into those showrooms. Meet the salespeople as well as the manager. Let them know you're focusing on the high-end consumer and would appreciate referrals. In return for the business they send your way, you will do the best you can to seal any deals between the customer and the wholesaler.

For the wholesaler, the benefit of working with you is that a professional plumber, one of his customers, is doing the installation and will warranty the product in the event of a problem.

By using your discount to purchase the products the customer's cost will be less and a level will be reached which protects everyone's best interest. In addition to the reduced cost, the customer will have the full warranty support of the installing plumber and the wholesaler stands less chance of that customer taking his business elsewhere. In addition, wholesalers will depend on you to buy other items related to the installation of these products. You are as important to a wholesaler as your customers are to you.

You benefit by having your mark up on the items you provide through this channel. Here's a good example: let's say you get a 30% discount on a $700 KWC kitchen faucet set. You will pay $490 for the faucet. You put your 25% mark up on it (Divide your cost by a .75 for gross margin). Your selling price to the homeowner is $653.33. Your profit (if there are no complications) is $163.00. You saved the homeowner almost $50! They would have paid full list price at the wholesaler in most cases. The homeowner is happy and protected and you have made more money.

Another way to make a similar arrangement is to give some power to the salesperson showing and quoting to your customers. You could arrange for him/her to tell the prospective customers that if they buy the product through you and you install it, they will get a 10% discount off the trade or list price and will also get the advantage of having the installing plumber warranty the item. If you know your discount is approximately 30% on most items you can easily gain by doing this. Ask your sales person to double-check your discount before you quote that discount to the customer. Make sure the salesperson keeps you informed about any changes in the discount you receive.

Get to know the floor sales staff well. You share the goal of wanting to make a sale and of keeping the customer from buying from someone else. Working as a team helps both of you. Make the sales staff aware of your flexible availability and make sure they have your rate sheet. They will be able provide the customer with a competent plumber for the job and you will have helped the showroom salesperson secure the order.

From the homeowner's point of view, she would rather hire a plumber who has been recommended by the people in the showroom, than use one she knows nothing about.

Home Center Managers

The manager at your local home center can help your business in various ways. Get to know him.

Homeowners often go to home centers to select fixtures, and then ask either the store manager or the manager of the section if they can suggest a professional to install the fixtures. There's your referral.

The reason the average homeowner goes into a home center to buy a product which he knows he cannot install himself does it for one reason; to cut out your profit. Because of this many contractors will not install these items. I think this policy causes them to miss many opportunities.

Make it quite clear to your client that he or she is responsible for any products bought at a home center. You cannot guarantee anything related to the performance of these items. Simply explain that if there is anything wrong with one of these items, you will do what you need to do and can do about the situation, but you will still have to be paid for all of the work time involved. The customer will also have to contact the home center in order to deal with any warranty issues on the products.

Once you are working in a home and have established an honest and trusting relationship with the homeowners, let them know that you can help from the start the next time a project is being planned. Assure your customers that you can also provide the same products they might buy in a home center for the same price, plus you provide peace of mind. They will be happy to know that a professional is taking full responsibility for the products.

I have had experience along these lines. A lady having trouble finding someone who would install a hot water dispenser she had bought at a home center called me. I told her that I would be glad to install it for her. I explained that since she supplied it she would be responsible if anything was wrong with it. I also told her that I would completely guarantee the installation of the insta-hot for one year, but of course could not warranty the product itself. She asked my rate and how long I thought it would take to install the hot water dispenser. After a brief conversation the job was scheduled.

I arrived at the job to find that she had bought an opened box item off of the discount rack. (I'm thinking, "Oh boy, here we go.") She asked if I thought anything was wrong with it and if all the parts were there. It looked okay from the outside but I told her we would have to hook it up and see. It leaked like crazy as soon as the water was turned on. Water was spraying out of the dispenser. It was obvious I would be unable to repair it. I just capped it all off and gave it to her to return. She of course paid me for my time.

I gave her some time to cool down and then I let her know I would help her out. I knocked a little bit of time off the invoice, considering I would be back to do the work. I also told her that in the future if she would contact me and let me know what she was going to do I could probably save her a lot of stress. I told her that with my discounts at the plumbing wholesalers I can usually (not all the time, but most of the time) get anything the home centers have at about the same price. I told her that if I had supplied that item not only would I have been totally responsible for it, but I also would not have charged for any extra work associated with a defective product.

Eventually she asked me to price, supply and install some bathroom fixtures she had chosen. She accepted my quote. I got the whole job and consequently a referral.

In instances like this, go ahead, do the work and make your money. Protect yourself and let the customer know how you can protect their interest in the future.

Getting back to the home center managers, there is something you can do which will help both of you. Many times home centers will change or discontinue stock. When they do they're glad to unload it at pennies on the dollar! Let the manager know you would liked to be informed about these items and if you can use them, you would be happy to buy them. The managers will be glad to contact you and let you get these things off their hands. You'll be able to purchase items you can resell for huge profits! You can pick up things like new faucets, stems or other parts, toilets, Whirlpool tubs, just about anything.

What if you got an assortment of brass stems for $20? Most of them would have a street value of up to $100 each! If you picked up a couple dozen or more at that price, you have boosted your future profits by thousands!

Supply House Managers

The managers of your favorite supply houses are also great sources for finding bargains. Much like the home center managers, they often are stuck with inventory that they want to get rid of.

For example, they have floor models that are changed regularly but nonetheless are brand new items. These are great places to find deals on high-end fixtures!

They also have something called dead inventory. When an item does not move for a period of time it is devalued to the branch and eventually begins to cost money to keep. You can buy these items for pennies on the dollar in most cases. The managers will love to show you what they have and will be very willing to negotiate it into your truck!

Noland is a good example. If they have inventory that is considered "dead" the branch is charged by the corporation for those items. If another branch finds out about the items and wants them, the first branch has to pay shipping costs, which can be expensive. This motivates the managers of all branches to avoid dead inventory and also to shop other branches for dead inventory that will move well in their market. They can buy those items at the devalued rate from other branches and not have to pay the freight. Of course this raises their margins and accomplishes the corporate office goal of keeping margins high and avoiding paying for things to sit around getting dusty. The whole process has a name. It is called **G**ross **I**nventory **M**argin **R**eturn **O**n **I**nvestment. Also know as GIMROI.

This process is not widely known outside of the wholesale business. Now that you understand how it works, you're in a position to work with your wholesalers to both make you a higher profit and to help them out as well.

As you develop your relationships with wholesale managers, you will also improve your chances to get referrals from them to the high-end customers that shop their showrooms. Make sure the managers know you want that business and that whenever you can you will help them move dead inventory through those referrals.

These managers are important people for you to get to know. You cannot develop these relationships overnight. It will take some time and effort but it will be well worth it in the long run.

Chapter 6
Profit Boosters

This part of the book will be the most fun. We are now going to go into great detail on about how simple it will be for you to make a lot more money on every service call.

This chapter will get your brain working on a different level. You will come up with even more great profit boosting ideas than we will give you here. You should have no problem increasing your profit by at least 30%. Many of you will double it almost instantly.

By definition a "profit booster" is any product you can sell or service you can perform that is above and beyond what you initially came to do and that will make you more money than you had initially planned to make.

Motivation

Of course you're motivated to make a profit while doing your job. You chose plumbing because you are ready, willing and able to understand the homeowner's needs and solve them. At the same time you can make a great profit.

As you approach each job, think about what opportunities you have to boost the profits on that job. What type of service or item may help enrich these people's lives? If it's a relatively new house, maybe they have some of the first generation 1.6 gallon per flush toilets and would like to get rid of them.

I do not mean to imply that you should find ways to talk people into things they do not need or want. Honesty should always come first. Be receptive to the improvements your customers might find useful and go from there. The satisfaction of your customers is your first priority and the source of your profits.

When a customer asks you a question, don't be too quick to answer. Get into the habit of trying to ask another related question first or at least thinking about it for a minute. You want to give the best and most accurate answers so that you don't have to correct yourself later. A bit of careful consideration, or further "research" often will lead you to a different conclusion and save you time and effort on a job.

Let's say you go to fix the water heater. As soon as you get in the door the homeowner asks you what you think is wrong with it. You say, "probably just the lower element." After you get into it you find out that it has two bad elements and a bad thermostat. What do you think the homeowner will think when you break this new news to him?

The better way to handle it would have been to ask what the problem was, how their hot water was acting and to say something along the lines of, "There are several things it could be. I'll need to check out everything and then I'll know what's wrong." In the "Up-Sell" section coming up, we are going to go into detail about how to ask the right questions.

If you're a multi-truck fleet, you have to motivate your employees. We all know how hard it is attracting and keeping good employees. Motivating good employees to work harder in order for your company to make more money is even more difficult. Your employees are your only connection to your customers. A disgruntled employee will cost you more money than any mistake you can make.

Why should your people work for you instead of the other big company down the street? What do you do for them that the other companies do not? Is working for your company a rewarding or fun job?

Motivated people who like their job work harder and are much less likely to steal from you. They care about the reputation of the company and feel like they're part of a team. Build this team spirit. Think creatively! You'll find all kinds of little things to boost your morale.

Run a contest with a prize like a dinner for two at a nice restaurant, or 2 days off with full pay. Maybe a weekend get away, or new tools. Tickets to the big game or race can be very impressive. It's true that not everyone will respond to this but many will and that will boost your profits. It's easy to get your people motivated if you push the right buttons. Remember, they need it and they deserve it.

The contest could be whoever brings in the most receipts over a certain amount of money in the time period wins or it could be the one who brings in the most new customers wins. Be creative. Reward your best employees. They deserve it.

Suppose once a month you bought dinner for two at the local steak house for the "Employee of the Month". It may cost you $50 as a business meal. You can add that to your expenses for tax purposes and you get several guys fighting for it regularly! Your company will profit and once the program is under way the competitive nature of most plumbers should more than take care of the rest. If you run a program and after a while it starts to get a bit stale, change it. Keep the competition fresh and your people interested in it and you keep the level of performance up.

Anything you can do to make working for you enjoyable and personally rewarding to your employees will boost morale, motivate your people and increase your margins considerably.

Spiffing

Spiffing is a great way to motivate your employees to become more productive with their selling techniques. The word spiff is an acronym for **S**ales **P**rogram **I**ncentive **F**und. Essentially, spiffing is rewarding your people with a bonus for selling certain goods or services.

For the multi-truck operation a Spiff program is an excellent tool. You want your people to make more money for you. You want your people to WANT to make more money for you. The best way to do this is for you and the employees to make more money simultaneously. You're turning your employees into partners who want your company to succeed.

The Spiff program can use many different things as the bonus, including days off, lunches, food items, products, golf outings and, of course, cold hard cash.

You can have a program that will reward everyone who hits a certain (attainable) gross receipts number in September and October with a free turkey for Thanksgiving. You can run one that says for every garbage disposer they sell in whatever time period you set up they will get $10 cash bonus. Now think about that for a minute. You buy Badger 5 Garbage disposers from the wholesaler for $55 each. You sell them off of your truck for $85 each, giving you a GP of $30. The actual numbers may be different in your area or for your company. It doesn't do you any good to have your guys just carrying these disposers around. By splitting the profit in a spiff you encourage the employee to go out and find new homes for these valuable products.

Apply this kind of program to other valuable products like water heaters and watch what happens to your bottom line!

The side benefit of doing business this way is you are going to retain your good employees. They are not going to move to another company for a few cents more per hour. They know that they, as well as you, will benefit from these profits. They'll want to keep it that way. Also, you're going to attract other high quality employees from your competition as they hear about the great way you're treating your people.

Attracting these excellent employees also gives you the ability to weed out the non-productive employees (if you still have any left) by being able to replace them quickly.

Getting sales training for your staff is an essential part of your Spiff program. It's important that your employees learn about the program and that they understand which sales techniques work without intimidating customers or being dishonest. If you're not comfortable providing this kind of training, hire it out to someone who is. There are companies that can provide sales training seminars, which will suit your needs and help you get the results you want.

Up-Selling

"Good things come to those who wait, but only the things left over by those who hustle."
- President Abraham Lincoln

Selling up is an art form. There is nothing wrong or dishonest in trying and succeeding in finding out where people have problems or would like to upgrade a certain item. If nothing else it can set you up for a future service call. Remind your customer you can fix other things for them right then and there, while you're working in their home. The homeowner may not be ready for the purchase or the repair right now, but they will call you when they are.

Have you ever left a meeting or appointment and thought "I wish I had asked about this or that"? Of course you have. This is what we want to avoid. The up-sell is a result of asking the right questions! This is how the one truck plumber gets rich, and the fleet company even richer! Get your customers to "remember" to ask you or tell you those things.

If you just ask this simple question, "How's your water heater?" you will be surprised at how much extra work you will get. Sure, most people will say something like "Oh, It's fine." But others will say something like, "Now that you mention it, I have been running out of hot water lately."

When that happens you have just identified a possible problem you can solve while you're at that customer's home. If there is a problem and you fix it, you have saved the homeowner the cost of another service call at a later date and you have made more money on that job by selling an extra service and a part or two more than you had planned on.

While you're talking to the homeowner about the water heater, ask how old it is. The U.S. Department of Energy statistics show that 30% of water heaters fail before their 7th birthday. That's a lot of water heaters. If you know how old the heater is and how well it's been maintained you have a good base of knowledge. Ask the owner if he has flushed it regularly. Most people have no idea that they're supposed to flush them out and no idea how to do it. If it hasn't been flushed in many years, it's best to leave it alone unless you want to spend hours trying to get the drain valve to pass enough water. Then it will probably not hold shut once you're done. If it has been flushed regularly then it will have a much greater chance of a long life, but this will be the exception.

Once the customer is talking about the heater, you will find out more things about its performance. There may be elaborate hot water rules about who can bathe and when. One parent may have to get up early in the morning to shower and give the heater enough time to recover so the other parent and the kids can have a hot shower before they are off to school and work. This kind of thing indicates the possible need for a larger unit or other means of supplying the hot water. You need to let the homeowners know there are other ways to get that hot water. Tell them you can help by recommending a new unit, which will meet their needs. Suggest a new larger water heater or going to a tankless water heater. We'll talk about those later. For now, let's get into some of the best profit boosters on the market.

FloodMaster

Here's a great up-selling opportunity based on a water heater accessory. The FloodMaster is a water heater leak detection device that works by using a small sensor attached to the discharge pipe from the T&P valve. It sits in the bottom of the drain pan required by most codes, or on the floor. These drain pans hold about ¾" of water before they start to drain from a small leak but this switch will activate on about 1/8" of water when the relief valve discharges on temperature, or if the tank ruptures. When the switch is activated it instantly shuts a solenoid valve in the cold water supply line, cutting off the water to the tank. The alarm sounds warning the homeowner that there is a problem.

This does two things. It shuts off the street pressure from the tank and it air locks the tank so if there is a catastrophic failure, the drain pan can do its job. Even if a large section of the tank fails, without the water supply on, the tank will drain like a giant milk jug turned upside down as the water glugs out at a rate that the pan can handle. Without the FloodMaster, the tank would dump its entire contents into the pan at once, causing an overflow and a flood.

Let's say you get a call to replace a water heater. You pick up a 50-gallon electric hot water heater from the supply house and get to the home to do the replacement. While you're getting everything together and working on draining down the old tank, find out how old it is, and if it had problems in the past. As you are cutting the old one loose, ask the homeowners if they have ever known someone who has had a water heater burst and flood the house or apartment. Of course the answer is always yes!

Most of the time they will go into great detail about how big a problem it was for that family, or how the family was displaced for weeks while their house was dried out and repairs made. If they don't tell you something like that, ask them, "Wow, did it do much damage to the house?" Listen carefully to what they say about that event and then let them know you can help them avoid that happening to them.

Say something like, "You might be interested to know there's a device available now which can be installed on your water heater. It will shut off your water in case of a leak and will prevent flooding.

I keep one on the truck, I'll be glad to show it to you if you think it's something you might like." Hand them a little piece of information about the FloodMaster to look over, The information fliers are made for your customers and are at the counter of most supply houses. While you're working on tearing out the old one you'll get the chance to find out if they'd like this item put in while you're there. You might add, "It costs less than your insurance deductible and if you would like it, I can put it in while I'm here so it's like getting the labor free. It really doesn't take long to install it."

Explain to the homeowner that the family will never have to worry about a catastrophic flood displacing them. Many times even when insurance does pay for all of the damage, the family is stuck living away from home for weeks while the repairs are made and the home dried out.

If your customer wants it you will increase your profit on that job by as much as $100 for doing almost no more work at all! Not only that, but you have sold that customer a valuable product which will protect the home for years to come. Put your company decal on the heater or on the FloodMaster valve so that when it actually does go off, you will probably be the one who gets the service call.

When the customer asks how much it costs, explain that it is reusable from tank to tank as they come and go, and it can be reset. That way if the problem that causes it to shut down the water is minor, you will come out, make the repair and re-set the FloodMaster, no problem.

The FloodMaster can also be used for washing machines, or any other item with a 3/4" water line and a possibility of leaking. Just put the valve in the line and the sensor where you need it.

There are many people who will want this type of item if you make them aware that it exists and that you can put it in. Make sure you carry the FloodMaster on your truck at all times. There is an old saying that you can't sell from an empty wagon. It applies to plumbing too!

Garbage Disposers

The common garbage disposer is an overlooked profit booster. You see one every day. Any type of repair might put you in contact with one. While you're on a job, check it out quickly. If you were called to replace it then this is not an up-sell but many times you'll be working on the kitchen faucet, fixing a leak in the drain piping below, hooking up a new dishwasher or any other repair that will have you in the area of the garbage disposer.

Look at the disposer and try to determine its age. Some 1/3 or 1/2 hp older models are quite large. Look for evidence of a leaking, cracked or broken chamber. Operate it to see that it works smoothly. Notice how loud it is.

If you've checked it out and found a small leak, you can point this out to the homeowner. "Mr. Homeowner, I've found another problem under here. It looks like your disposal is starting to leak. It's not too bad now, but it will get worse and you may want to consider replacing it before it causes water damage or mold growth in your cabinets. If you want to stick your head under here I'll show you what I'm talking about." (Always offer to show him, even if the problem you find is under the house. This lets him know that you are not trying to con him and lets him know he's really the one in charge.)

Once he understands he might ask if you have one. Of course you do, and your response will be, "Yes sir, I always keep at least two on the truck because of just this kind of thing. Here's the good news. Since we found it now, you're saving quite a bit of money by not having me or another plumber make a separate trip out at a later date to replace it. The price of the disposer is $85 (or whatever your sell price is) and my company will guarantee it completely for the first year. Would you like me to go ahead and replace it while I'm at it?" (Always ask for clear instruction even when you already have implied consent to make a repair. Never do anything without it.)

That was an easy one. Let's say you check it out and it still works, but it is noisy, big and looks like it's about 10 years old.

You're working away and hearing all about his kid's sports prowess or the neighbor's cat digging in the flowerbed or whatever and you say, "by the way, how old is this garbage disposer?" The answer will be something like**,** "I don't really know, I've had the house about 10 years and it was here when I bought it. Is there something wrong with it?"

Ask, "Does it still work ok? Is it loud?" Respond to the answers with "The normal life expectancy of a household appliance is about 8 years. Right now it's not leaking and it's still working so I don't know, it might run another 8 years or might die tomorrow. If you are interested, I keep a new on the truck that is the same power, half the size, much quieter and under warranty. If you want me to install it while I'm here you won't have to pay someone to make a special trip later when it does die."

Many times you can start this conversation without ever checking out the disposer. Maybe it was already disconnected for some other reason like a new sink or counter top install. Either way, talk to the homeowner about it and watch what happens.

You'll be shocked at how many disposers you'll sell this way, and you're doing the homeowner a great service by helping him avoid future repairs. You're replacing an outdated appliance with a new, modern one under warranty. This is a valuable service and you are entitled to profit from it.

Low Flow Toilets

Who would have thought, a few years ago, right after the Department of Energy made it mandatory that all toilets flush on 1.6 gallons, that the huge hassle would turn into an overflow of opportunity?

Here's how it happened. Starting in 1992 the United States Department of Energy mandated that all residential toilets operate on only 1.6 gallons per flush. That part is not the big news. The big news is that most of the toilet manufacturers did not have time to do all the research and development work necessary to make the toilets they were producing function properly.

Everyone knew that toilets could flush on 1.6 gallons because commercial toilets with flush valves had been doing it for years but the fixture companies just were not ready to do this with residential toilets.

Many of the first generation low flow toilets were just gerri-rigged versions of the same toilets these manufacturers had been making for years. Some of them used a plastic "dam" in the tank to measure out the water, some used a revised flapper or other flush system, but the toilet itself was still wanting the 3.5 gallons of water that it was designed to operate with. The bowls and the trapways were the same as before.

This caused a backlash from the American public, demanding that the government ban these fixtures and recall the law. The government did neither. Many people went about converting the first generation 1.6's back to their 3.5 form by cutting out the dams or replacing the weird flappers with the old style parts which were still readily available. Many plumbers even make these "Adjustments" to satisfy their customers.

It took about 4 or 5 years for the word to get out about the success of the Toto brand of toilets from Japan. The American manufacturers had done all their R&D work and had begun to produce 1.6GPF toilets that worked. But the damage to the 1.6GPF toilet's reputation was already done. Many homeowners were still living with toilets they hated but could do relatively little about. Hundreds of thousands of these homes are out there!

Some people even began going to salvage yards to buy older toilets in order to replace these poorly working new models. The older toilets were in such demand that it even became expensive to buy them at the salvage yards.

Many homeowners are not aware that, for a small investment, their hard to keep clean, first generation low flow toilet can be replaced with one that works great. Homes built after 1992 have a good chance of having poor quality low flow toilets in them. Even now, some of the builder grade toilets do a poor job in handling waste with the 1.6GPF amount of water.

Right now, in your area, there are hundreds of thousands of opportunities for you to help homeowners solve one of their most annoying problems. When you're working in a house built after 1992, ask, " Do you have the new low flow toilets in this house? Do they work properly?"

Many homeowners will go on for some time about how they hate their toilets. They will tell you stories about them stopping up or having to flush repeatedly. Here is the opportunity to set up a light renovation job with high profit potential.

Tell customers that those toilets are outdated and now there are many brands available that do work and will function properly. As you know, Kohler, American Standard and Toto all make very good 1.6GPF models. There are other acceptable models made by Mansfield, Briggs and some other lower end manufacturers.

Explain that replacing the toilets they use most would reduce their water bills. They will also save time and energy by not needing to scour the bowl after each use. Tell them that replacing the round front models with an elongated front will add to their comfort and they're easier to keep clean. Offer to recommend a good model for them and provide them with a firm quote for replacing one or more of them in the bathrooms they use most.

After they tell you how many they would like you to price for them, follow through on it the next day with your quote. Chances are you'll have another profitable job on the schedule and another very satisfied customer whom you can ask for a referral.

While we're at it, how many calls would get if you ran an ad in a local paper that simply said something like; "Tired of stopped up toilets? Get a new toilet that works! Call XYZ Plumbing Today 555-1234"

Shower Heads

The 2.5GPM showerhead is another profit opportunity. While you are talking to the homeowner about things in the house, ask him if the house has low flow showerheads and if so, how is the performance? Most people who have these showerheads don't like them. Many people simply drill out the orifice or take out the flow restrictor, causing them to run out of hot water faster and use more water than they really need. Money is wasted because too much water is heated and used.

There are solid brass chrome plated showerheads on the market, which put out a very forceful and impressive flow at 2.5GPM. They look a bit smaller than the normal head. It seems as if there's a lot more water coming out of them than there really is, due to the velocity.

These showerheads are available at home centers and at many supply houses for anywhere from $8-$12. You can sell them off your truck for $20 by simply doing the following. Once the homeowner has told you she has a need for such a product, say, "I'll tell you what. I think I have something that'll solve this problem. I'll be right back."

Go out to the truck and get one of these heads. Show it to the customer and ask to see one of her showers. Run the shower as it is first. "Yes ma'am that is a poor flow. Let's try this one." Now swap out the heads and run the shower again. "How do you like that shower?"

She will love it of course and will want to know the cost. Tell her and add it to your invoice while you're on the call. You solved their problem and again, boosted your profit on the job for almost no additional work.

As another up-sell option and even better profit booster try the Moen Revolution showerhead. The Moen is more expensive but is a high quality product that acts as a showerhead and shower massage unit. They're well made, very attractive and are also impressive performers. Your higher end homeowner will love it at first sight.

Try the ad tactic mentioned in the low flow toilet section. Your phone would ring off the hook with an ad like this; "No More Wimpy Showers! Call XYZ Plumbing today and start tomorrow refreshed 555-1234,"

Faucets

If you're doing some work for realtors or landlords, always keep a couple of builder grade faucets on your truck. A decent Delta lavatory faucet available to you for $30 will sell off your truck for $60. You can get a two-handle kitchen sink faucet for about $45 and sell it for $90. Here is how this happens.

You'll get the call from a realtor to look at a sink and/or faucet that needs some work. Maybe it's dripping or just draining slowly. While you're there if you notice that the faucet is in poor condition, suggest to the realtor or homeowner that the best option would be to replace the old rusty faucet with a new one. Add that small improvements like that help houses sell more quickly. You're ready with the faucet on your truck. You have an instant profit booster.

You have also helped the realtor who will remember you as a problem solver looking out for the interests of her customers. Realtors are busy people. If you are their "resource" they'll want to use you on a regular basis. Ask the realtor to include your card or your information flier in the package of information given to homeowners at the closing. This allows you to indirectly contact hundreds of potential new customers each year!

Dealing with a realtor in a good part of town is a great way to reach the customers you want without spending a lot of money.

Extended Warranties

How many times have you bought a piece of equipment or appliance at a large store and been sold an extended warranty? Now think about that for a minute. Why do you think these stores offer them? Because they are huge profit boosters!

You know that the U.S. Department of Energy statistics show that 30% of all water heaters fail by their 7th birthday. You know it's unlikely that an electric tank type water heater will need repair within the first 5 years. All the water heaters on the market have a five-year tank and a one year parts warranty at a minimum. You also know that these warranties only cover the parts to be replaced. Offering a service warranty to go with the heater when you install it is your opportunity to increase your profits exponentially.

You can offer the warranty for gas models also. You'll want to structure it so that the warranty gives the purchaser protection to pay your labor in the event that the heater needs a repair during this time period. It should cover all aspects of the repair including the service call fee (if any), and all required parts and labor during the warranty period.

Most homeowners will purchase your warranty on the spot. You'll sell the warranty at a rate equal to the cost of a full hour service call to make a repair on the water heater. The overwhelming majority of the heaters you install will never require this service within the warranty period. When one comes along that does, your homeowner will be very glad she purchased the warranty.

Here is a sample of what your warranty should look like. Review it, have your lawyer look it over also, Make any changes you need to protect your interest. Add or remove any details of the warranty to fit your business. The more detail you have in the warranty the better protection it is for you. The more information you have listed on the warranty the easier it will be to verify, when and if made, that it is the same piece of equipment you installed.

M. Scott Gregg

"Your Company Name Here" Appliance Limited Service Warranty*

The Appliance listed below shall be covered under this warranty for a period of ____ years from the date of installation to the original purchaser of the appliance.

Appliance Type: _____
Model Number: _____
Serial Number: _____
Date of installation: _____
Installing technician: _____
Signature of
Installing Technician: _____

Items covered under this warranty are any and all ancillary parts, labor charges and fees associated with making any repair or replacement of this appliance within the warranty period.

Not included in this warranty is damage due to misuse, natural causes or disasters, floods, lightning, water damage, power surges, pressure spikes or over-pressure conditions, fire, or impact damage from external forces or any other cause of interference with the operation of the unit from any other source. This warranty is applicable to the original purchaser only and is non-transferable.

*This warranty is an example only and not intended for actual use. No permission is given or implied for its use. Your warranty should be reviewed by your lawyer for completeness, exposure and accuracy before using it in the course of your business.

Maintenance Plans

Another great profit booster that is often overlooked is the ability to sell a service plan along with a piece of equipment. You will need to decide what equipment will carry a service plan and draft the plan to specifically cover the items you want to cover, define what work will be done and what the limits to the plan (if any) will be.

You could prepare one for water heaters of course, but also for any other equipment you install. If your company will do any boiler work, make sure that all of the preventative maintenance work necessary is done on the boiler to avoid breakdowns. Avoiding after hour emergency calls is a great way to boost your profit. I also recommend that if you are going to be doing the P.M. work on water heaters, figure the cost of and install a ball type drain valve with a hose connection on the new tank. Do this at the time of installation to make blowing down or draining the tank faster and easier. This will save you lots of time in the long run and will get you in and out of the homeowner's house with a lot less trouble for both of you. Figure the cost of this work in with your price of the maintenance plan.

Part of your service should be to remove and inspect the anode rod of the water heater for wear and replace it if needed. To do this you may need an impact wrench to keep from having to fight the nut. Anode rods are usually installed by machine at great torque and will not be easily removed with hand tools. Of course it can be done by hand but the impact wrench makes things a lot easier,

Many people, including plumbers, don't really understand the function of the anode rod in a water heater. It's not there to protect against bad water. It's there to protect against stray electrical currents. Static electricity and stray currents are everywhere in today's homes, due to computers, multiple televisions and other devices. Good information about the anode is available on the State Water Heater web site and on the Bock water heater web site. Here are a few websites to help you fully understand the function of the anode rod:

http://www.hotwater.com/htmlbulletins/bul43.htm

http://www.rheem.com/Documents/ResourceLibrary/TSB_Common/1201.pdf

http://www.gguldens.org/anodes.html

http://www.chilipepperapp.com/GWH.htm

If the anode rod is being eaten away within the first year or so, check the electrical connections to the heater, especially the ground. Recommend the homeowner have an electrician check the house ground as well.

Another way to protect the water heater and ensure you are not going to be replacing the anode rod regularly is to add dielectric unions to the water connections at the time of installation. This keeps many of the stray currents that get into the piping of a home from getting to your water heater. If plastic water lines such as CPVC or PEX are used of course, this is not necessary.

Hot Water Recirculating Systems

This is one of the most overlooked areas of profit boosting and problem solving for plumbers everywhere. Most homes have at least a ¾" hot water main running in the crawl space all the way to the far end of the home. Very few have a recirculating system.

Most systems that mount under a cabinet and use the cold water pipe as the recirculating line can do an adequate job of keeping the hot water line hot. They typically provide warm water at the farthest outlet and hot water at most other locations. Laing makes many of these systems and they work well with tank style water heaters.

With today's larger homes, there is typically a very long delay between the time a faucet is turned on and the time when the hot water is ready to use. The low flow fixtures add to this delay by restricting the flow to tiny amounts. The result is that new homeowners are dissatisfied with their hot water delivery. Some blame the hot water heater when, in reality, it's doing all it can do. The cause of the problem is the combination of longer piping runs and lower flow fixtures. This is a good opportunity for the enterprising plumbing contractor to sell a very effective hot water recirculating system.

If you are a repair or remodeling plumbing contractor, you can go under most of these homes and find a very easy solution to the problem. The lag time can be greatly reduced by installing a hot water recirculating line from the end of the hot water main back to cold water line that feeds the water heater. You will need to have an expansion tank, an aquastat like the Honeywell L6006A or L6006C and a pump.

If a tankless water heater is the hot water source, a small storage tank will help in the piping by keeping the pump on longer and off longer. The small storage tank will also reduce the likelihood of a cold water sandwich, meaning a short burst of cold water goes through the heat exchanger. When this happens the homeowner notices a short temperature fluctuation in the hot water piping. If she's in the shower it is not welcome!

A system like this can be installed in a few hours and at a great profit. You will be able to make as much profit on one job as most people make in a week. You'll also be saving the homeowner thousands of gallons of wasted water per year! In most cases the time lag between when the faucet is activated and when the hot water is received will be reduced to 10 – 30 seconds instead of well over a minute or more.

Laing makes some very good under sink models, which are available at most plumbing supply houses. They're self-contained and hook up beneath the farthest sink. They are best used in smaller homes or those with tank type water heaters. Only a few models work well with tankless water heaters. You'll need to do some homework in order to find them.

Another company that makes a good product for retrofit is ACT Metlund. Their system uses one or more buttons hidden beneath the countertop. When the homeowner wants the hot water to be instant, she pushes the button. The pump will then run until the built-in thermostat adjusts. Then the pump is off until the next time the button is pushed. This saves even more energy by not running a recirculating system at all unless the homeowner wants it to run.

If you are a plumbing contractor who does new construction as well as remodeling and repair, then you are sitting on the best profit potential ever!

Very few builders will pay to have you install a recirculating system in a new home. Here is an opportunity to make a huge profit, solve the homeowner's problem and make a customer for life.

While your crew is piping the house have them install one more 1/2" water line from the hot water line of the farthest fixture in the uppermost portion of the home all the way back to the crawl space or water heater. If it's a large home, run a line from the farthest point at each end. You only have to run the lines to a point where you can get to them easily later. In some homes this may require a bit more piping but on most you'll only have to drop the lines into the crawl space. At the end of the line add a valve and cap. Figure out where the pump will be and how the system will hook up at a later date. With an outlet strategically placed your power wiring can be done to most small pumps with an appliance cord wired into it and plugged into the outlet.

You might even want to slip the electrician a little something to drop an extra outlet near the location you want your pump in order to avoid any wiring issues later. We've all seen many things done for the price of a case of beer or a lunch out.

All of this will most likely cost you less than $20 per home. Once the neighborhood is finished you simply have your service department make up fliers letting the homeowners in that neighborhood know that you can install a hot water recirculating system. Let them know that it will save them thousands of gallons a year in wasted water and get them instant hot water at each fixture.

Since you already know the home and what's there you can even include an estimate of the cost. No surprises for the homeowner! A properly designed flier should get your phone ringing off the hook. She already knows the cost and at this point all you have to do is schedule the job and get it done.

You have already run most of the piping and now all you or your team have to do is add the proper equipment, hook it up and get paid for the whole project.

Your price is highly competitive. Other plumbers may give an estimate for the same job, but they won't be aware of the pre-run piping. Their estimate will be higher than yours if they're able to do it at all.

Word of mouth is powerful. Even though you've distributed fliers, not everyone will take the time to look at them. Once you have put in a few systems, however, the neighbors will hear about you. They'll find out about your prices and your work and you'll hear from them.

Home Plumbing Inspections and Warranties for Realtors

Home inspections have been done on home sales for years. People who have a strong basic knowledge of homes and construction but do not specialize in a trade, much less plumbing, usually do them. You could offer a home plumbing inspection and warranty to go along with the other protections offered for an existing home's sale.

Your inspection should provide a complete report. Offer a warranty if no problems are found. If there is something that needs attention, tell the homeowner that if it's repaired by you you will be able to provide a warranty. You don't have to call it a warranty, you can call it a service agreement. The service agreement or warranty will offer to protect the homeowner from any surprises and you'll be able to make the profit up front on the calls that may or may not need to be made later. Here is how it works to boost profits.

For your inspection fee you will inspect the home for plumbing issues and make any recommendation for repairs that can be done by you or someone else. This will take an hour to two so a charge for two hours' time will be sufficient. You're not going to make any repairs unless directed to do so by the homeowner after the inspection is complete. You'll be paid for the labor to make the inspection. This is not a profit booster at this point unless you found repairs to be made and are asked to make them. Many times you can do the minor repairs while you're there.

Now you have made the inspection and know that all of the system works properly. You know that the water heater is in reasonably good shape and have checked the toilet flappers and fill valves for bad seals and replaced them if necessary. You have checked all the faucets and showers for leaks and drips. You've checked the garbage disposer and kitchen sink sprayer and made sure the sink drained properly. You have been under the home and looked over the drain system. You've checked all the hose bibs for function and looked for any frozen pipes that may have just been valved off. Test every fixture as if you had just installed it for function and leaks. Fill them to the overflow with water and drain them, watching for any leaks or malfunction above and below the entire time.

Once all this is done it's reasonably safe to say that everything in this home will be fine for at least two years. There will be some homes that do have something happen during that period and you'll have to fix those things. Most will not have any problems. You can sell your service agreement or warranty with this in mind and for the cost of a two hour service call and another $25-$50 or so for parts your service agreements will quickly add up to a nice profit.

For example, let's say you've done 10 inspections for $200 each. On five of them you made repairs as directed by the homeowner based on discoveries from the inspection and boosted your profit a bit. We will assume that the other five opted out of the service agreement because you already found and repaired the existing problems. We'll average it out to a $75 dollar profit on each repair.

10 Inspections @ $200 =	$2000
5 repairs @ $75=	$375
5 Service agreements @ $350	$1750
Total profit	$4125!

All of this profit is up front. This money is sitting in your account or paying your bills now. Time spent based on 10 inspections at 2 hours each for the above profit is 20 hours total. That works out to a bit over $200/hr! Not half bad.

Let's say that during the service agreement period one of these homes has a minor problem. We'll call it a fill valve in a toilet. You're in and out in an hour including the drive time. The homeowner has signed a no-charge invoice for the full value of the work. (You need this for tax purposes of course as it counts against your profit as a loss.)

You may have had to give up $100 of that profit you already made in raw cost (material and labor time) but that money has been in your possession working for you for a long time.

Some of these little surprises can be a lot more expensive. That's the value for the homeowner. Once you're doing these more and your agreements are numbering in the hundreds instead of tens, watch the profits grow to larger numbers. Even with several costly warranty repairs made by you, your profits will remain extremely high.

The homeowners who needed the repairs have benefited greatly. You'll be able to tell the next homeowner about the client who, "just last week" got a new water heater installed without cost under his service agreement. "It would have cost him $500! No problem. One phone call and we were out there." The customers who have a repair covered by your program become good references for you. Ask for a letter of recommendation explaining the service and the cost savings they received.

Become a Specialist

There are many places where profit margins are higher than the usual repair or replacement market. Take that opportunity to differentiate yourself from others in the business. Customers will tend to gravitate to someone who markets himself or herself as a specialist.

This could have easily been a separate chapter in itself but the point is that specializing enables you to boost your profit by charging more for your knowledge and concentrating your work on things you know very well or are most comfortable with or just enjoy doing. Any time you can pick the work you want to do, you will be better off both financially and emotionally. Capitalize on your strengths. Specialize.

Water Heaters

If your marketing strategy has you positioning yourself as a water heater expert, most of your calls will be related to repairs, replacements, or perceived hot water problems. There are many different ways to make hot water. No one way is right for everything. There are tankless coils in boilers, indirect water heaters, gas, oil and electric tank type heaters. The tankless water heaters in gas and electric are rapidly gaining in popularity.

The problems can be just as wide ranging. There are problems of overheated hot water, running out of hot water, burst tanks, inconsistent temperatures and many others. Most plumbers have a very good working knowledge of many types of water heaters but very few "specialize" in water heaters.

If you take time to learn about these systems you'll be able to provide a very valuable service to your customers. You'll increase your profits by being the specialist in these areas.

Often just having the right ads in place with this information will get you the first call. Include a line in your ad that says "Hot Water Heater Specialists. We diagnose and repair all hot water related problems". Customers will notice that and will be more likely to call you because you are a specialist. If you are going to go this route you need to deliver on your claim.

Delivering on your claim is easy but will take a time investment on your part. All water heater and boiler manufacturers run special training classes to increase knowledge of their products. Ask your supply house managers about these classes and let them know you're seriously interested in getting this type of training. The classes are usually free and only require that you show up ready to participate and learn. The classes are lively and fun. Some even include hands on training.

Every class you take will increase your knowledge. You'll learn from the other people in the class as well as from the class itself. Even if you just went to a class on gas water heater repair last week and another manufacturer is having one this week go to it. You'll learn something about their heater that you didn't know or maybe something about their competitor's water heater that you didn't know.

Some of the manufacturers, especially the tankless water heater manufacturers or the boiler manufacturers will have much more in depth classes. These pieces of equipment are quite a bit more involved and tend to have much more advanced technology than the average tank type water heater. Many have diagnostic capabilities that can help you determine problems and even view fault histories of the unit without needing an advanced computer degree. Most of the time the diagnostics are flashing lights or number fault codes that can easily be used to find the source of a problem…if you know how to get to them.

In many areas of the country there are homes that have the tankless hot water coil in the boiler that supplies hot water to the home all year. These tend to have three types of problems.

The first is sometimes an inadequate flow rate for the hot water demand of the home. A coil that will only pass 3.5 GPM or less will not be able to run much more than one or two showers at a time before it either runs out of flow (a noticeable decrease in pressure at the fixture outlet), or it may allow colder water to move through the coil (a noticeable decrease in temperature at the fixture outlet).

Two of the common possible fixes for this are to either go to an indirect hot water heater and do away with the coil, or to install a flow restrictor in the hot water outlet to ensure that the flow rate of the coil cannot be exceeded.

The indirect hot water heater is the best overall option in that it will allow you to give the customer a large supply of quickly recovering hot water. Most indirect hot water heaters have a lifetime warranty and can produce hundreds of gallons of hot water per hour! They are available in many different sizes from a 30-gallon lowboy style to over a 100-gallon storage. Even the smaller 30 and 40 gallon indirect hot water heaters can produce well over 100 gallons per hour of hot water! That's a big improvement. By going with the indirect you can easily control the temperature of the outgoing hot water and the usage rate will not affect the delivery temperature when sized properly.

Temperature is another problem that many tankless coils have. Many homes have these coils in boilers that have a set point of 160° to 180°F. This means that without proper temperature regulation the domestic hot water to the home is as high as 180°! This dangerous condition is very common in the older neighborhoods. A water temperature that high will cause immediate scald burns and will cause great injury and even death in small children or older adults! It's amazing how many homes still have this problem. You can easily fix this by installing a mixing valve to temper the hot water to a safe level. Almost every homeowner with this problem would gladly pay you to solve it if they only knew it could be fixed. All you have to do is get the word out.

The sad thing is that most homeowners are not aware of this. Most have never been told that they can have this problem corrected. Look out for this in every home you go into. If you find a home with this condition immediately speak to the homeowner. Let them know you can provide a safe hot water system for them.

This is an easy up-sell of your job. If you were smart and have a mixing valve riding along on your truck you can fix the problem while you're there. The customer will appreciate the fact that the problem is solved and another service call is avoided.

Make a note of the address and send a direct mailer to every home in the neighborhood, explaining the situation and how you can solve the problem for them. Since most homes in these neighborhoods are the same, you could give an estimate right on the mailer if you like, or just encourage them to call for an estimate. If a direct mailer is too much work, just make up fliers. Use your other customer as a reference (with their permission of course) and drop the fliers in every mailbox.

The last problem that these coils tend to have is a flow restriction caused by mineral deposits plugging up the coil. This can cause under heating of the hot water as well as a poor flow rate. The signal for this is the customer who says he used to have plenty of hot water, but now he doesn't.

You may be able to clean the coil out with various solutions or you may need to replace the coil. Of course if you are a hot water specialist the customer will call you to do either job.

By being a "Hot Water Heater Specialist" you will get calls from people who have unique applications as well. Maybe they just need a small point of use water heater to a detached garage or possibly they would like to do away with their old large tank and get one of the new tankless water heaters. Some will call just to have you relocate their existing tank to another location or to remove and reinstall it for a floor renovation.

Tankless Water Heaters

This one could have been included in the "Water Heater" section above but is important enough to stand alone. The market for tankless water heaters is the fastest growing market in our trade. In fact it is doubling or tripling each year! Someone has to be prepared to do all that work.

The impact that tankless water heaters is making on our industry hasn't been matched since indoor toilets! Nothing in my lifetime has come along to revolutionize plumbing like these amazing machines.

Many people who want a tankless water heater look in the phone book to decide whom to call. If you are listed as a "Tankless Water Heater Expert" or at least reference them in your ad, you're likely to get the call.

There are many people who have paid tens of thousands of dollars for bathroom renovations only to find out that their hot water capacity is not adequate. They cannot fill the tub or use their new high flow shower system for more than a few minutes before running out of hot water. This is a problem that can best be solved by a high flow rate tankless water heater.

One of the largest tankless water heater manufacturers (Noritz) has a commercial grade unit that can put out 13.2 gallons per minute out of one heater. That's 752 gallons per hour! The unit is the size of a computer and mounts on the wall. They can also be installed as a modular system with lead/lag capability in multiples of up to 24 units giving a mind blowing 317 GPM or 19,020 gallons per hour! All without taking up floor space and being able save huge amounts of energy by not having to maintain a stored volume of hot water. This is the best solution to a large flow hot water need for the customer.

Noritz and Rennai both have less expensive residential models that can be installed together for large capacity needs. Tankless water heaters like this come with two main challenges, sizing the gas line and venting the unit.

The gas piping must be sized correctly for the heater or it will not function properly. You will only get small amounts of hot water before the unit will trip out on low gas. The easiest way to avoid this problem is to run a dedicated gas line from the gas meter to the unit. The meter should also be checked for capacity, as with a large unit the Gas Company may need to upsize the service or meter or if it's an LP job a larger tank may need to be installed depending on the size of the existing LP tank.

There are several manufacturers of the new, coated flexible gas piping systems that are perfect for this job. You can run the line without joints back to the meter, hook it all up and be done in no time with plenty of gas feeding the new tankless water heater. Often you can hook them up where the drip leg at the meter is by adding a tee and putting the old drip leg back in at the bottom.

The venting of the units is pretty straightforward. Most require a special stainless steel vent though and B-vent cannot be used. Some have vent systems that are proprietary and are sold with the units. Others have to be bought separately. They all have vent length restrictions for horizontal sidewall venting and most also have restrictions for vertical vent length. Be sure to read the manufacturer's instructions carefully. Don't let the venting scare you off though. It is all pretty simple once you have one under your belt. Once you know how they go in, they're very similar to a "B-vent" or "Snaplock" venting job.

Locating the tankless water heater is part of the fun. Due to their size they can be installed in many places including the attic, crawl space and garage. Some models can even be installed outside the home. This of course frees up the most space for your customer. No other kind of hot water heater has this kind of installation flexibility.

When they're installed in a garage in order to replace a tank type water heater, the homeowner is able to remove the stand and the pipe bollard that the old water heater had, providing more usable space. Of course you can remove these while you are there if you choose.

Specializing in these tankless water heaters can and will be a big profit booster. Many plumbers are just not familiar with the things that these remarkable machines can do and are not comfortable dealing with them. This is a great opportunity for the plumber who will seek the knowledge and do the work.

There is a plumbing company in California that got so interested in tankless water heaters that they put up a sales center in a strip mall. They call it "Advanced Water Heaters". They have every unit of the Noritz line in it. (That happens to be their favorite line.) The units are working units, hooked up to showers and jets so prospective buyers can see them operate. This business now does only tankless water heater change-outs! They do about 40 every month and have grown to several trucks.

As more and more of these units are installed there will be a market for plumbers who are trained in the service and repair of them as well. The manufacturer's service training will get you into a position to do this work quickly and you will have a high value service to add to the things you can offer your customers.

Many plumbers who are called to look at these heaters will remove the cover and see the various wiring harnesses and computer board and put the cover right back on and leave. They will not have the knowledge to make the repair and (hopefully) wisely tell the homeowner to call someone else. If your ad in the phone book lists you as a tankless water heater expert, you will quickly get that call.

Now here is one of the best reasons to sell and install a tankless water heater. If you get a 35% mark-up on a water heater, a tank type water heater, which you sell and install, gives you a lot less gross profit than the tankless. Look at the table below.

Tankless cost	$850	Tank cost	$250
Sell $ @ 35%	$1307	Sell $ @ 35%	$385
Gross Profit	**$457**	Gross Profit	$134

That is an increase of 70% in your gross profit and that is just for the equipment! Remember that company that is doing 40 every month.

Radiant Floor Heat

Radiant floor heat offers many opportunities to make money. There are very few people who do this kind of work. It is an area of specialization. Those who do the work and do it well are making a nice profit margin and for good reason.

A radiant floor heating system, whether it's for a bathroom or a whole house, gives the most even comfortable heat available. It's not drafty, dusty or dry. It feels like indoor sunshine. Radiant is a not only a very efficient form of heat; it also is efficient economically.

When Radiant jobs are done incorrectly, there is a risk of uneven heat or overheated floors occurring. There is also the danger of warping hardwood and/or cracked concrete or tile. Possibly the system can fail altogether.

There are companies that sell Radiant floor heating systems and run training classes for all levels of experience. Find out about them through your supply houses and get the training needed to get into this type of business. Usually these training classes and seminars are free. The manufacturer sometimes provides hotels and meals for out of town attendees.

As you develop a relationship with the manufacturer's representative you will find out about many more classes and seminars that will increase your knowledge about their products and systems. Taking these classes and seminars is a great way to boost your profits by enhancing your knowledge and making you more prepared than the average contractor to install and service these systems.

There are many ways of installing radiant heat in existing homes. For instance, the pipes can be installed in concrete slabs or in the joist bays beneath the floor. They can be installed under tile in a mud base or in one of several above floor engineered board systems. They can be installed in a lightweight concrete over-pour or in a sidewalk or driveway as snow and ice melt systems. You can wrap tile shower walls or large tubs in the tubing to provide warmth. They can be installed under all known floor coverings including tile, laminate, hardwood floors, vinyl and carpet.

The more you know about these systems the more likely you are to be the one who gets the job. Homeowners who are familiar with radiant heat know the value of these systems and will pay top dollar for them to be installed properly.

Hi-Velocity HVAC Systems

Hi Velocity HVAC systems have been around since the 70's, but they're still relatively new to our trade. These systems use a pressurized trunk duct and small (normally 2") insulated flexible ducts to distribute the flow of conditioned air to a space. They use a small air handing unit for the pressure and can have various types of heat sources added to them including heat pumps, electric strip heat or hydronic coils. They can also have electronic air cleaning devices installed into the system to help enhance or maintain indoor air quality. (IAQ)

I.A.Q. is fast becoming a booming business in itself. People are becoming more and more aware of the quality of the air in their homes and are paying to have elaborate systems installed to clean it. Watch as this new trade develops. If you can get the training, be ready to become a leader in your area and watch the people flock to you for your systems!

The hi-velocity HVAC systems are very good problem solvers in that they can be installed into spaces by "fishing" the flex ducts through walls, ceilings or floor cavities without harming the existing finish. In older homes with classic architecture or plaster ceilings and walls this is a great way to add air conditioning or supplemental heat to the home. The HVAC system has been marketed as "air conditioning for homes that can't be air conditioned."

The installation of the systems is quite a bit easier than a conventional system. Large trunk ducts are not needed. The equipment is more expensive but the added value to the home makes these systems very saleable.

Remember Kelleher? They're the company with the teddy bears. They have installed over 500 of these systems since the 70's and still install them today.

Again, free training is available for these products from the manufacturer. All you need to do is ask around and find it. The manufacturer's representatives are a great source for the information, training and design help for these systems as well.

Geothermal HVAC

Geothermal (much like radiant heat) is a specialized system. These systems use a series of piping loops buried below ground to use the earth's steady temperature to provide most of the energy to heat or cool a property.

Geothermal systems are able to provide the heating or cooling at a substantial energy savings over other systems because most of the energy required to add or remove heat from the water comes naturally from the ground. These systems promise substantial savings to owners over all other systems even though electricity is the main energy used.

Training is available from the manufacturers of these systems for those who want to be able to size, install or work on them. Plumbers who get into and specialize in these systems are able to attract plenty of high profit business.

Total Re-pipe work

Another good thing to specialize in is the total house re-pipe either for water or sanitary purposes. Many people with older homes have problems with failing copper pipes. Also, homes built with Quest piping that is coming apart need this service.

If you are a specialist in this field you can get the job done faster and better and be paid for your expertise. The work is not difficult. All you have to do is fish PEX or CPVC around to replace what is there.

Ask the customer about other water issues he may have before you get started. You may find a better way to run the system, be able to solve more problems, and make yourself more money at the same time.

You may find homes that have old cast iron with problems, or a building sewer with a root problem. There is only one way to solve root problems. The piping must be dug up and replaced in order to replace the bad portion letting the roots in. Usually this takes the better part of a day including the filling of the ditch and patching of the yard. You must get to the root of the problem!

These jobs are a bit messier, but high profit. The customer gets a long-term solution which raises the value of his home considerably. It's a good deal all around.

Water Treatment Systems

Why would you want to re-pipe a home with bad water lines and not suggest that the homeowner have the water tested? The number of missed opportunities to sell and install water treatment equipment in the USA is staggering, probably well into the millions of dollars a year! The only reason for this lost opportunity is not many plumbers have taken the time to get the little bit of knowledge it takes to be able to spot and up-sell to solve these unique problems for their customers.

In 2004, in Montgomery County and Prince George's County Maryland 5,500 homes had to have their copper pipes replaced because of pinhole leaks. That is 5,500 homes that NEED a water treatment system! Suppose the average cost of each of those systems is $3000. That's $16,500,000 in available, and much needed business there for the taking!

A home that has copper piping that has been eaten away by corrosive water needs to have the water problem addressed even if you are going to re-pipe the home with a plastic piping system.

Hard water and water that has a low pH (acidic water) are harmful to every single appliance in the home. The water heater, clothes washer, refrigerator, dishwasher, disposer and even the drain lines can be harmed by these problems. You might even find out that the same house has had the thin metal traps leak before and the homeowner had to pay for those repairs.

You have the advantage of being in your customer's house while he is in need. He is looking to you to solve problems. Every water treatment equipment company that sells through plumbing wholesalers will be glad to give you the training you need for these jobs.

In one of these classes not only become a trained installer, but you will also learn how to go out and sell the product for a very nice profit. You will make a lot more money at 30%-35% selling a water treatment system that cost you $1500 and solves your customer's problem that selling them another water heater at $250 in a few years.

A simple water test will get you started. You will most likely sell the repair or re-pipe job and add a nice profit for installing a water treatment system!

Even a hot water heater replacement can lead you to these jobs. You will be in a home doing a replacement and talking to the homeowner who will remark that they just don't make water heaters like they used to. They may say something like, "This thing is only 5 years old and it's already leaking."

This is your chance to do some digging. Talk to the homeowner about how water quality can change. Take a sample. Explain how water quality affects every appliance that uses water, often shortening the life of those products greatly if there is a problem. Here is how the conversation might go:

You: "Boy these copper lines are in bad shape. It looks like you'll need a re-pipe. I can fix the leaks you have but most likely you will have more any day and they will keep showing up until you replace the piping."

Homeowner: "I was afraid of that. We had some leaks last year and they told us the same thing."

You: "Having plumbers in to fix repeated leaks will eventually cost you a lot more than going ahead and replacing them. How about I give you a quote for the job?"

Homeowner: "Ok, when could you schedule the work and should we have you patch the leaks we have now or just do the whole thing at once?"

You: "I'll tell you what. I'll give you a quote for the total re-pipe and if you want us to do the job for you, you can consider this visit your free estimate. I can get you on the schedule for next week and if you just catch the drips until then you'll save some money by not paying to repair something that will get torn out in a week."

Homeowner: "That's nice of you to do."

You: "You know that having your piping eaten up like this is a sure sign of a water quality problem. The same water that's eating your water lines is also probably eating away at your drainage piping and every appliance you have hooked up to it, like the dishwasher, clothes washer or refrigerator. Anything that touches this water could have its effective life shortened by quite a bit. I really would recommend considering a treatment system that would protect your home from more problems like this in the future. Most of the time a treatment system costs less than replacing a couple appliances and a lot less than repeated plumbing visits for patch jobs.

I can take a water sample and get it to the company that supplies the treatment equipment we install. It won't cost you a thing and I'll be able to know what's causing all this damage. I'll give you a separate quote on a treatment system that will solve the problem once and for all.

No cost or obligation up front for you, I just want you to know what the problem is and that we can solve it for you if you like."

Homeowner: "Sure go ahead. How long does it take to get the results back?"

You: "I'll probably know something first of next week. When we get back together, I'll go over the report with you and bring you some information on whatever equipment is recommended. Most of the time we can hide the equipment so you don't have to see it yet it will be accessible. You'll also get some other advantages, like better tasting water and cleaner, longer lasting clothes.

You might get comments on bad tasting water or other appliances that have had problems. These all help you make your point and the sale.

You have already established a good relationship with the homeowner by offering free estimates and no charge for the visit if he gives you the re-pipe job. Sure, it's a gamble that he could go shopping around for the re-pipe, but look what you've already done for him. You've handled the problem in a very professional manner and given great advice. You're providing, at no charge, a full water report and another estimate for the treatment system. If the timing works out, you'll probably be told to install the new system at the same time you do the re-pipe.

Customers appreciate these low or no pressure sales tactics. They are smart enough to know that you make money on these things and they are smart enough to know that they have a problem that needs solving.

You will get these jobs because you'll be the first one who wants to solve the problem once and for all and you represented yourself or your company with the professionalism that your customer deserves.

Once you have gotten permission to test the water, you'll send it off to the water treatment company. You'll receive a comprehensive report with a recommendation that will advise you on how to provide your customer with high quality water. This will protect the entire plumbing system and every appliance hooked up to it.

Once you've reviewed the report, make a follow up appointment with the homeowner to give him a copy and go over it with him. Explain why he needs the system and how it will pay for itself by protecting the plumbing system of the home and avoid other costly repairs later. Bring some manufacturer's literature on the recommended equipment. Be prepared to go over what regular maintenance will be required, if any, and have your quote ready.

No you won't get every job for your trouble, but you'll be surprised at how many you do get and the profit you will make for being the one who was able to meet that need.

You will probably also get some very nice referrals to neighbors who have the same water problems! Don't forget to leave a several of your cards and ask the homeowners to refer their neighbors to you. Remember that, "Ask and ye shall receive" applies especially to referrals!

Water Filters

Similar in benefits to the water treatment systems but even easier to sell are water filter systems. Inline water filters are available from many companies. They remove odors, particles, sediment and iron from water, making it taste and smell better.

Water filters are smaller than softeners and fit in easily in just about any location. They come in types and sizes, ranging from whole house cartridge type systems to under counter drinking water or behind the refrigerator models. The last two are for dedicated use at points where drinking water is to be the only item filtered.

What makes these products easy to sell is the fact that you're there, working on the plumbing system and have access to the water. Your clues to the homeowners' need for these products are easy to spot. If you notice the sinks or toilets are stained with iron, especially around the bathroom drains, you know there's an iron problem with the water. If the water smells bad, or you find out that toilet fill valves need frequent repair, you may have particles in the water that should be filtered out.

Information on water filter systems is available at your favorite plumbing supply house. The people there will help you get in contact with the manufacturer's representative who can get you all the training and information you need. By doing this you'll be ready to sell these items when you see the need, and again, boost your profits considerably.

Disposal Charge for Removed Equipment

Another great profit booster is the disposal charge for removed equipment. Many times a homeowner will ask if you can haul away an item like an old toilet, disposal or water heater. Charge for it. Think about this. Once you write up the invoice you can no longer be paid for any time spent on the job. You will have to use your time to go by the local dump to dispose of the item or items. It is your gas, your vehicle and your time. You should be compensated for it.

So what is this disposal fee worth? Of course it's worth exactly the amount your customer is willing to pay and no less than you are willing to accept for the service. You will have to feel your way around this one to determine the amount. If the dump is on your way home and doesn't charge you to dispose of equipment, then this can be a fairly low rate.

A homeowner is not likely to want to pay very much for you to throw away a garbage disposer. If you get a $10 Disposal fee you can drop it in the trash at your office and you made an extra $10.

Ask, "Do you have a trashcan I can put the old disposal in or would you like me to dispose of it for you?" If she says she would like you to dispose of it, tell her your disposal fee. Explain that it includes the extra stop and time at the dump. If she says that's fine, you've added another $10 to your profit by doing almost nothing.

Larger pieces of equipment are a bit different. Most people don't have a pickup truck to haul away an old hot water heater or toilet. Not to mention that they would much rather not have to struggle with these heavy (or nasty) items. They're quite willing to pay you to dispose of these things. Twenty to fifty dollars, more for larger items, is a reasonable disposal fee. In other words, explain the options and let the homeowner decide. In most cases he or she would much rather have you take care of the disposal and spare them the heinous task. You make it sound so easy and reasonable!

Just wait until you have to replace a length of sewer pipe under the house! You can just about get anything you want to remove that, but keep it honest. Your goal is to always be paid for the services you perform, not to take advantage of anyone.

The Biggest Profit Booster of All

The absolute best profit booster you will ever find is one simple question to ask each customer on every job. You must ask this question BEFORE you begin to write up the invoice. And you must ask it on EVERY job!

"Is there anything else that you would like me to look at while I'm here?"

If you haven't asked this question, the customer may say, as you hand him or her the invoice, "Oh yes, one more thing before you go." Or "I just have one question about my this or that before you go." This will frequently result in you spending from several minutes to an hour making another repair or doing work of some kind or talking to the homeowner for free when you should be "on the clock".

It is easier and much more useful to ask the question yourself up front while the meter is still running. Then, if the customer has another issue you can solve, you have the chance to fix it on the spot and you're still on their time.

Customers appreciate your caring enough to ask if there is anything else they would like you to do. This lets them feel in charge and reminds them that you are working for them and you are done when they say you are done.

Frequently you will get to perform a minor repair or adjustment on some other piece of the property while you're there. Of course you will charge for it and the homeowner is pleased because the repair has been made and it won't entail another service call.

This important question will automatically add to your profit margin and help you avoid having to hear, "Would you mind taking a look at this before you go".

Chapter 7
Time Savers, Gadgets and Specialty Tools

Gadget or Gold Mine?

There are many items available for tradesmen to help them do their work. Many of these items are worthwhile investments and some are things that look useful in the package but will rarely if ever actually be used in the field.

The purpose of this section is to pass along suggestions for items that will either help you do a better job, do the same job in less time or just provide an alternative way of getting something done.

There are some items that are so simple many homeowners have them. They're worth keeping on your truck. Others are items that many homeowners do not have and would never buy. Many of these items are what separates the weekend DIY'er from you, the professional plumber. They help you get a routine job done efficiently and will keep you from fighting your way through jobs that would take a homeowner hours to do.

Any item you find that makes doing your job easier, better or faster is worth considering. Every time you use one of them you save time, which makes you more money.

Good, high quality tools and equipment that are not abused last a very long time and quickly recover your initial investment.

The proper tool or piece of equipment for a job is essential to your success.

Here is one of the best pieces of advice in this book! You plumbers who have been doing it for a while already know this and you have already read it once but it is worth repeating. You people new to the trade may not know just how important it is. Never skimp or buy cheap tools or equipment! You get what you pay for, always.

You must, however, weigh the cost of the item against how often you will use it and what the benefit of having it really is. If it's too bulky to carry or you find you avoid going back to the truck to get it, it is a waste of money no matter how cool it looks on the shelf or how good it sounds in the ad.

The toilet lift comes to mind here. It's a neat gadget that looks like a small duct jack or engine hoist. Its purpose is to lift and hold the toilet to position it above the bolts. One you have it lined up you lower the toilet with it straight down, avoiding the usual strain associated with this task. All that sounds and looks great. But why in the world would you want to make the extra trip? By the time you get it, assemble it and fool with it you could have set the toilet and hooked it up and probably even tested it out.

I'm sure there are some folks who can use the toilet lift do strength constraints but once you need it, it's time to get a good helper with a strong back or consider looking for another line of work. We'll stick to the gadgets that will actually get used and make your life a lot easier.

Not worth it you say. Think about this. If you save just 5 minutes every work day, you will save over 20 hours per year! With a little ingenuity you can easily save 3 or four times that much. What will you do with 60 or more, free hours per year?

Drain Stick

We'll start simple. The Drain Stick is a little device that is made of nylon and is about 20" long. It has a flat handle and has about a ¾" end that is slightly angled with a little comb like end on it for grabbing hair.

This is found at the home centers and can be stuck into many drains like a lavatory or shower drain instead of dragging out the drain machine to remove the hair clogs normally found closest to the point of entry into the drain.

Stick it in, give it a spin or two and pull out all the gooey hair. It is amazing how many hair clogs you can fix with this simple little tool. No power needed, or any disassembly of the drain piping. It grabs all of it. You typically have to cut off the hair it brings out! I have seen the other barbed looking drain sticks but they get hung up on everything. Look around to find the long skinny one described here. Every plumber should have this device on his or her truck.

Lavatory Nut Tool

Here is one you make yourself that you may not have thought of. It won't cost you a penny and will save you a lot of time. How many times have you strained, using a basing wrench, trying to tighten the faucet retaining nuts that hold the faucet on the lavatory? Or, how many times have you broken one trying to remove it? A lot I'm sure.

Most all of these nuts are the same. Most have two or four tabs or wings to grip them and do not do well with a basin wrench that tends to break off the tabs.

Take a short piece of 1-1/2" PVC pipe and cut the end to catch these wings. Four ¼" notches should do it at the opposite sides (north south east and west) and you will never struggle again. Even removing old stuck nuts will be a breeze!

A Water Key

There are many times when you need to cut off the water to the home to make a repair. For example, that will be necessary any time you have to replace or repair a stop valve under a fixture that is not holding. There are only two ways to cut off water to the home. You either find the main shutoff valve (normally under the house) or you shut off the water at the meter down in the meter box.

Getting to and shutting off the meter valve is the easiest and fastest way to do this. You won't have to get dirty crawling under a house. The down side is, it's only easy if you have a water key. This is a long "T" handle tool that allows you to reach into the box and turn the flat valve handle with little effort and without exposing your hand to the nasty biting creatures who can make the valve box their home.

By using a water key and this valve you also avoid problems that may be associated with the main valve. Many times when you shut off the home's main water valve it allows water to leak past, or it can stick closed, causing you to have another thing to fix under the house. By using the valve in the water meter box you don't have to trust the home's main water valve. It's much faster to pop the lid on the water meter box and shut this valve than it is to put on coveralls, go under the house, shut off the valve, take off your coveralls and go back in to make the repair. You will save at least 5 minutes using it on these jobs and probably a lot more on average.

EZ Snap Closet Bolts

How many ways have you found to cut off the closet bolts after you set the toilet and find the bolts are too long for the china caps? A bunch I'll bet.

All of the tiny hacksaws and little saws, which hold the hack saw blade only, involve you standing on your head beside a toilet hacking away at a bolt while you bang up your knuckles and waste valuable time.

If EZ snap closet bolts are on your truck you can avoid this. These bolts simply have a recess that allows you to use a crescent wrench to break off the last ¼" or so of the bolt when they are too long, without harming the threads.

There is absolutely no reason to ever buy any other kind of closet bolt sets. There is hardly any difference in price. You'll be selling them off of your truck for $3-$5 or more, and getting all of your profit. In addition, the homeowner is not paying you to mess around with those silly little saws for ten minutes while you try to cut off those other bolts.

Possibly the best reason for using these bolts is that very likely you will be the next guy to have to pull that toilet, for whatever reason, and you will know that the threads are good.

Bucket Boss

I got one of these things a few years ago as a Christmas gift. My mother saw it at a home center and thought it would make a good gift for me. To be honest, I almost took it back. To me it looked like a yard sale candidate. Just for kicks I tried it out.

Was I ever glad I did! This bucket boss allowed me to set up a bucket with the tools I normally use on a service call and leave the heavier stuff in the truck. I can toss the items for that particular job in the bucket, carry them in and get to work.

Add a piece of carpet to the bottom of your bucket to protect hardwood flooring. When you're under a sink you'll really appreciate having everything you need right there waiting for you. No running back to the truck!

The hard part is finding a bucket with a heavy enough handle for our trade. They are out there. Look for the buckets for wall joint compound, paint, cement products or other heavy items. These buckets have a handle that can hold up to years worth of heavy tools.

Dewalt Cordless Tools

The Dewalt cordless tools have really come on strong. They are high quality and frequently outlive the abuse you think will kill them.

A good example of this is the one I bought when building my house. My plan was to add screws to the floors after the framers had used a nail gun to put down the sub floor. I was on my own framing crew as well, so once the rest of the guys left I could go around and put screws into the floor.

I bought a 14V with two packs, a charger and a hard case. I figured it would burn up during the job at some point. I put down 15,000 1-1/2" drywall screws into that floor and I still use the tool today! It has been over six years as of this writing!

The packs got so hot you could not hang on to them. I would have one charging while I used the other. I know that is a bit more abuse then it should get, but that's how it went. It is still one of my favorite tools and gets used almost daily.

I would recommend the new 18V series for those buying new cordless items. That power pack is used more often than the smaller versions and the 18V model has noticeably more power.

You will be able to use lanterns for under the house, saw-z-alls, skill saws, regular drills, hammer drills, radios, wet vacs and more. If they all use the same pack then you can interchange them as needed and keep a good supply of fresh packs on hand at all times.

Anything that keeps you from dragging out a cord and searching an unfamiliar home to find an outlet is a great thing. Many older homes only have the old non-grounded two hole plugs which are very inconvenient. I know there are adapters, but if you have a cordless version of that tool you will save a lot of time. The absolute

BEST thing that Dewalt has done for us plumbers is come out with that little jewel of a wet vac!

Now when you first think about it a 2-gallon wet vac is not that great. After all, for mopping up a real mess or a large clean up job 2 gallons is just not a useful storage amount. That may be true, but consider this. When you have to replace a fill valve and you turn off the water and flush out the tank, the toilet only holds about a quart of water.

Even if you have to pull the toilet for whatever reason, the 2-gallon capacity is more than enough to suck the remaining water out of the tank and the bowl. Plus you still have capacity to tidy up after the job is done! All without needing to drag out a cord or carry in a larger machine.

The handle allows you to carry the Dewalt along with your other tools without worrying about it coming apart on your way up the stairs. It's small enough and light enough to take anywhere.

Every plumber in the country should put Dewalt on his or her Christmas card list just to thank the company for coming up with this tool! The first week you use it you will save so much time that it will make up the hundred bucks or so that you spent on it.

Think of all the fill valve replacements you do in a year. If you don't have this machine on your truck, go buy one tomorrow and send me a Christmas card for telling you about it. Come to think of it, Dewalt ought to send me one too, for the same reason.

Nextel

This is a bit above gadget status but Nextel is fast becoming a "must have" for just about every trade person. The Direct Connect feature that Nextel has allows instant contact with other Nextel users. Most supply houses, inspectors and other businesses you deal with are using this technology.

It does not sound like much on the surface but having this type of communication will save you hours each year. You will no longer have to wait hours for responses that can be handled in minutes.

At this writing, the only problem you will have with Nextel is the possibility of poor signal in remote or mountainous areas. If these are your areas, it may not be the best choice. Find out first.

With the power of instant communication comes the responsibility to be polite and considerate of other people. As a bonus to this book here is a lesson in Nextel etiquette.

Always remember to turn your phone onto the "vibrate" feature when you're at a meeting. It's embarrassing as well as rude for someone to key up the phone to talk to you and expect that it's a convenient time for you. It's disruptive to any meeting or anything else you may be doing.

Do not be the person who makes that call either. Having direct contact allows you just that, so you need a way to politely say, "Excuse me, can you talk?" The same button, which lets you be rude, also allows you to be polite. Every time you key up, the other party gets a "beep beep" to let them know someone wants to talk. That is all you need to do. Just key the button and give the other party the opportunity to respond. If there's no response it's safe to assume the other person is not able to talk. Try again later.

If you receive a "beep beep" and cannot talk, if possible let the other party know you will have to get back to him or her. This only takes a few seconds and keeps the other party from wondering if you have stepped away from your phone.

The "Alert" feature is handy but using it for every contact becomes obnoxious. It's best to reserve this feature for times when your call is urgent.

There are times when you will prefer to have your conversations remain private. Turning the speaker "off" makes this an easy and polite way to do so.

Nextel phones are not CB radios! Keying up with a nice loud "How 'boutcha Bob!" is not proper use of the tool and makes the user look less professional.

The Internet

The Internet has become a primary source for plumbers. Not only is it good for doing basic research but you can also get all the information you need on just about any product there.

Manufacturer's websites have cut sheets, rough in data, specifications and performance of almost all equipment listed. From these sites you can usually find contact information for their local representatives, a source for training and any other information you may need.

Using e-mail for business is also a preferred method of sending important documents. You can request product quotes and availability before you go out, and they will be there when you get back. Print what you need, leaving the rest on your computer or saved to a disc, thereby taking up less space and using less ink.

Another great use of the Internet is information sharing. There are various sites you can go to post or answer questions relating to all types of products and installations. Some examples are:

plbg.com

This is my personal favorite. On this site you can share information with other plumbers from around the world. Sign up is free and you can give or get advice on just about any plumbing related and some HVAC questions. The knowledge base at this site is second to none and the people are all there to help.

radiantpanelassociation.org

This is a great free site with a bulletin board. It's dedicated to the radiant heating part of the trade and is a great resource for those interested in learning more about radiant heat or in finding a solution to a problem.

Heatinghelp.com

Dan Holohan Associates runs this site. Mr. Holohan is well known as a guru of heating knowledge. His organization and this site are both great resources for you to learn more about hydronic heating, steam heating and radiant floor heating. Some questions posted on "The Wall" make it into Plumbing
& Mechanical and PM/Engineer magazines. This is how well respected this site is within the trade!

serviceroundtable.com

This is a pay site where your membership of $25/month gets you into either the plumbing roundtable or HVAC roundtable part of the site. For $50 you can join both.
This site is a good resource and should return dividends on the investment of membership.

Chapter 8
Records and Bookkeeping

Records and bookkeeping are the necessary evils of business. You have to be able to keep track of where and who your customers and suppliers are, as well as any transactions you have had with them. It's also important to keep track of your other contacts, such as realtors and contractors.

Of course, keeping track of your cash flow is crucial. Cash flow will start from the very minute you buy your first tool or part. Keep a complete record of everything coming in and going out for tax purposes as well as keeping you aware of how you stand financially at all times.

In this chapter we will look at several possible solutions to help you get started. The goal is to find an efficient way to keep your books, save time doing so and to prevent expensive mistakes in the long run. Much of your record keeping will have to be done in your free time, so you want to make sure you don't waste that precious time.

If you are in business for yourself you only have to work half days. You can pick any 12 hours you want.
-Unknown

Act! Contact Manager

One of your first challenges will be keeping track of all your contacts. Simply stated, everyone you meet or do business with from customers to suppliers to manufacturer's representatives, are contacts. You need to store their information in a way that is helpful to you and will help you keep track of who is who.

The best program I have found to do this is called ACT! It's made by Symantec, the same people who make the Norton antivirus software. It is like a bionic address book on steroids!

It's impossible, in this book, to list all the things Symantec can do, but we can hit the highlights. The description of the program here should serve as enough of a teaser to at least get you to consider it.

If you have never had a program like this you may not even know yet what managing your contacts means, but read on and you will see how you must be able to do this in order to be successful in any business.

As you build your business or career you will interact with hundreds and even thousands of people. Every one of them is important to some degree or another. How will you go about remembering all of these people over the many years of your career? You will need to not only remember their names, but their other contact information as well. (Phone, fax, address, etc.) It's helpful if there are little hints to remind of who they are, why they are important, where you met them, what kind of things they like, or any other information that is important about that person.

Once you start collecting business cards, you will soon find that you have them everywhere but that you can never find the one you need.

First, Act! acts like an address book. You enter the name, address, phone, fax, e-mail etc into the template information boxes for each contact. This builds your database of contacts. Enter everyone you know, including friends and family. Once they are in there, you can really start to have fun with it. You can change the basic information boxes if you want to personalize it. Add notes to the notes and history tab to remind you of important things about this person.

It will really surprise you when, after the first year, you notice just how many people you know and deal with. Once they're in there you can use the database to manage them as you need. You can use ACT! To make template letters, quotes, proposals, fax cover sheets or any other type of document that you use regularly while dealing with your contacts. It will save you valuable time when preparing a document. You can even use the mail merge feature to write one letter and have it written personally to every contact you choose. This makes direct mail or flier campaigns a snap!

Another way to group your contacts by a common characteristic is by using the group feature. This will allow you to assign a group name like Wholesalers, Customers, Realtors, Friends, Family, or any other category you need for your contacts. Then if you want to do something with these contacts as a group, two clicks of the mouse will do it. You can look up and narrow your database (temporarily) to just the people you want.

You can also do a lookup based on company, first name, last name, phone number, city, state or even zip code. The possibilities using this feature are only as limited as your imagination. You'll also be able to obtain histories and reports, which will provide relevant information on your contacts.

One of the most powerful features of ACT! is that you can keep track of every document or piece of information in reference to any contact as you print it. The information will be on that contact's entry, under the notes and histories tab, and you can set the program to do that automatically. This way you have a complete record of everything you've done with a contact for future reference. You will be able to see that you sent them a letter on 12/10/2004, what the subject was and even where the file is located so you can retrieve it and review it without hunting through your file cabinet for it.

You will be backing up the database every day to a CD or DVD. This way you are protected if your computer ever crashes. (I know that never happens right?) But if you are backed up to disc, just reload the program, go to the restore feature and your ACT! program is just as you left it.

If you are going to use this program, get and read the *Act! for Dummies* book. This book is listed in the recommended reading section near the back of this book and for good reason. It is very well written and reads easily. It will help you get the most from your program.

In fact, I will go further and recommend that you buy the book before you buy and install the program. This way, you will already have a basic knowledge of what it can do before you ever get started.

Once you get through the first couple of chapters you will be using your ACT! program well enough to make it worth the investment. By the time you get through the book, you will know all it can do for you and will have picked out the things you want to use right away. Later, if you want to do broadcast faxing or large mailings, having the book will be a great advantage.

Act for Dummies can also help you with another program, called WinFax Pro, which can work with Act! It enables you to send your faxes right from the screen you're on without having to print first.

As an example of how this feature is helpful let's say you want to prepare and send a quote to 10 different contacts. The quote will be the same for each so you set up a quote template, use the mail merge feature and do a look up to the selected contacts. Now you have done one document that will be printed and faxed individually to 10 different people but will look as if you have done it just for each one individually.

You will fax it using WinFax pro right from Act! With one click you have set the fax to go out to those 10 contacts. With one more click you print a hard copy for the file cabinet. You can walk away while you computer and modem do what would have tied up you or your secretary for over an hour! Pretty cool stuff huh?

Part of the Act! program is the "Side Act!" list feature that allows you make, save and re-use lists for any occasion. It can be a parts list, a safety checklist, a list of things to pack on your vacation, a grocery list or whatever else you need.

Act! also has a scheduling feature which has a calendar and will allow you to schedule tasks, meetings and calls for any contact. With this feature you'll be able to prioritize your appointments and set up alarms to remind you about them.

That's enough about Act! for now. If you use it, it will not take long for you to get dependent on it or for it to start saving you time.

Microsoft Excel

As you begin to set up your business you will need to be able to make up simple spreadsheets to help you with some of your paperwork. Excel is a great program. It will let you develop your material lists and inventories and even make price changes easily for your field documents. A spreadsheet is a ledger that will enable you to make any chart you need. It uses columns and cells. You can fill the cells with words, numbers, prices, or anything else you need.

Let's say you want a price list to use in the field for common toilet parts. You would set up a spreadsheet with the descriptions of the parts, the cost to you, the sales tax, the total cost and the sell price. This way, when you use an item on the job you have the prepared price ready to write onto the invoice for each part, without having to look it up every time. This makes your billing consistent. You can even add other columns for your use at the office, like the suppliers part number you will re-order from.

Here is an example of what a spreadsheet like this might look like.

M. Scott Gregg

Part Description	Unit	Tax	Price	Sell Price
Flapper for Flush valve-Korky Red	$2.39	$0.11	$2.50	$ 7.49
Flapper for Flush valve-Korky Black	$1.72	$0.08	$1.80	$ 5.39
Tank Float for ballcock	$0.35	$0.02	$0.37	$ 1.83
Toilet tank poly supply X12"	$0.42	$0.02	$0.44	$ 2.19
Toilet tank poly supply X20"	$0.53	$0.02	$0.55	$ 2.77
Standard wax ring	$0.27	$0.01	$0.28	$ 2.82
3/8" fip x 3/8" CH comp stop	$1.28	$0.06	$1.34	$ 5.35
EZ snap closet bolt set	$2.84	$0.13	$2.97	$ 5.94
China cap set (white)	$0.34	$0.02	$0.36	$ 1.07
Flush valve handle	$0.98	$0.04	$1.02	$ 2.05
Am Std flush valve seal (snap style)	$0.70	$0.03	$0.73	$ 4.39
Am Std flush valve seal (screw style)	$0.33	$0.01	$0.34	$ 2.07
Rod type tank ball	$1.06	$0.05	$1.11	$ 6.65
Fluid master flush valve	$5.88	$0.26	$6.14	$ 11.67
Fluid master repair kit	$0.83	$0.04	$0.87	$ 3.47
Ball cock repair kit - 4 screw type	$2.41	$0.11	$2.52	$ 5.04
Tank bolt and gasket set	$1.12	$0.05	$1.17	$ 2.34
Flapper type flush valve	$3.81	$0.17	$3.98	$ 7.96
Mansfield M-08 Ballcock	$5.63	$0.25	$5.88	$ 11.18
Mansfield M-08 repair kit	$1.13	$0.05	$1.18	$ 4.72
Mansfield style flush valve	$6.50	$0.29	$6.79	$ 13.58
Mansfield 210 Seal	$0.51	$0.02	$0.53	$ 4.26
Mansfield 208/209 Seal	$1.50	$0.07	$1.57	$ 9.41
Tank Spud Gasket	$2.11	$0.09	$2.20	$ 4.41

As you can see, this program can be very useful. What you cannot see is that only the raw cost of the items has been entered. Excel uses formulas that are pretty simple to do the math for you and fill in the other cells of the sheet.

Now I don't want to turn this into an Excel instruction manual, as there are plenty of other books out there that can help you use the program, but here is how to do some basic things with Excel.

The vertical cells are columns and the horizontal cells are rows. The individual spaces are called cells and anything in them is the "value". It can have words numbers or letters in them. The rows are numbered 1,2,3,etc, and the columns are A, B, C etc. Each cell is identified by its position by column and row such as A1, A2, A3 etc. If you enter a number as the value in a row or column, you can use other rows to manipulate it (do the math) for you.

Look back at the example spread sheet. The "Unit" column is where the raw cost is entered. This is column "B". Only type the things between the quotation marks, not the marks themselves. In the tax column to have your tax figured automatically you type in the first cell "=B*.045". (4-1/2% is the current sales tax rate in VA) You hit the enter key and the math is done. Now we do not have to do that for each cell. You can highlight the cell and right click to get to a sub menu and hit copy, then highlight the other cells you want to use this formula, right click on them and click paste. Or you can use the auto copy feature by placing your mouse arrow on the lower right corner of the cell, click and hold and drop down the column to copy that formula to every cell you want to use that formula. Once you have done that, your "B" column will all have the math done for sales tax!

To add the two together to get your total raw cost, enter into the first cell of the "D" column "=B+C" then hit enter. There is your math for that. Now drag that formula down your sheet like you did the tax and that is done.

To mark up your raw cost to get the sell price, you enter a formula like this "=PRODUCT(D1,1.3)" This puts 30% on the items. Drag that down like the other columns and you're finished. You can make the multiplier anything you want and change it from cell to cell if you need to.

Once this sheet is set up the way you like it as your costs change all you have to do is put in the new raw cost and all the other math is done automatically! This makes it easy to keep your cost current and your pricing in the field consistent as you get started.

There are of course many other ways to get and use field pricing and many other things you can use Excel for in your business, including making bid summary sheets.

For the summary sheet, just use two columns. In the first (Column "A") use the name of the line item and in the second (Column "B") put in your total price for that item. Once you have everything in, select the cell below the last item and use the "Quick Sum" button (The one that looks like a backward "E") and highlight all of the cells to be added. When you hit the enter button, you will get your total.

As you experiment with the program and learn more about it, you will use it to perform other functions that will help you get your paperwork done faster and with fewer errors.

Quickbooks Pro

To use this program effectively, and to take advantage of all it can do, you will need to get a good instructional manual on it. I suggest *Quickbooks-All in One Desk Reference for Dummies*.

You can do quite a bit, if not all, of your bookkeeping with Excel while you are getting started or while your business is mostly "side work". Once you are full time, however, you need to have a program that lets you manage your business and helps you keep track of your money flow.

Quickbooks lets you do all of this. There are many good reasons why it's the most used program by small businesses. It enables you to keep track of your entire inventory, quote projects, order materials, set up and track your accounts payable and set up and track your accounts receivable. It will help you identify overdue invoices. It will allow you to do your monthly accounting easily,

You will always know if you are making or losing money, and how much. At tax time you can just take the proper print outs to your accountant for easy filing.

Consider using this Quickbooks Pro as you start out so that your transition to full time will be that much smoother. Just as I recommend you do with the Act! book, get the Quickbooks Pro book before you buy the program. You will understand what the program can do and how this expensive but invaluable purchase can help you run your business.

Chapter 9
Philanthropy

True Charity Work for Plumbers

Much of this book has been devoted to money and the pursuit of it in the course of your business. Our focus has been making money, getting your money, controlling your money and making sure that you always get paid what you are worth . All of that is very important if you plan to have your business support you and your family. You cannot ask nicely to get your kids into college or to get food at the grocery store. You can't get your car worked on just because you are a nice guy unless you have a very personal relationship with your mechanic.

So, why is there a chapter on charity in this book? Simple. You are going to run into people who really do need your help and have little or no way to pay you. Hopefully, you will not run into too many of these calls but they do happen.

You will really want to help some of these people. You're going to make a lot of money in this business. Helping people who really need your help will make you feel you are really making a difference in the world. Guess what, you are!

Helping those who really need it is one of the greatest joys in life and should not be reserved for Christmas only. You will grow your soul just as fast as your bank accounts if you let your conscience lead you to do the right things at the right time.

We are talking about people who are really in need, not people who just refuse to work. The communication skills and people skills, which you've learned, will help you separate need from laziness.

There will be times when you'll arrive at the home of an elderly person or a mother on welfare whom you know has little or no money and has a great need to have a repair done. This is where you should allow the kindness of your heart to lead the way. Many of these folks have too much pride to accept your charity but there are ways to allow them to accept it without hurting their pride.

Older people can be especially easy to offend. Many have never taken a penny of help and don't want it now. Most don't need your help anyway, but you will run into those who do and it will be easier if you have planned for situations like this beforehand and are prepared.

You can get as creative as you like in how you go about discounting or donating your services. Be sure you understand that what you are doing is between you and that homeowner and there is no reward for your action. This act should be in the true form of a gift and kept to yourself, your spouse or bookkeeper. If an invoice has been created and you get it signed, you can use that as a write off on your books to lower your taxable income but that is as far as it should ever go.

Fix the problem of course and help out in any other way necessary. If you are going to try to write off the job as a loss, write up the invoice as you normally would. Have the client sign the invoice and ask him to write you a check or pay "X" amount, or to forget the whole thing, depending on what you decide. Once you have left the property, note on the invoice the amount charged or that it was donated and use that to adjust your books accordingly.

You are only going to get one chance to go through this world. The more people you can help out along the way, the better off you will be for it. You will be a happier and more complete person. Your riches will truly be those you cannot measure.

The most important "wealth" you can build is happiness, for you and for others. This is wealth that cannot be taken away or taxed and is more valuable than anything else in the world. It is a true blessing when you get the opportunity to help someone.

Favors for Friends

Once the word gets out that you are a plumber, you'll find many friends, or people who say they're your friends, will call you looking for free or discounted work. This is a touchy thing. The more you give away, the more your so-called friends may test your limits.

I have learned by experience. A good friend of mine assumed, because he and I had known each other so long, that he could extend my services through him to just about anyone he knew or was related to. I explained the difference between work and a hobby to him and that was resolved once and for all.

In a nutshell I told him, "You're my friend. I'm glad to help you out any way I can and you can always count on me but I have to draw the line somewhere. If I don't I will be spending all my time working for free for everyone all my friends know or are related to and before long, I'll be broke."

I encourage you to help out your friends and family, just be prepared to draw the line when the time comes. Draw it gracefully, without putting anyone in a difficult position.

I also found that there is no good way to discount your service to a friend or neighbor. It seems as if you're in a situation where you are damned if you do and damned if you don't. It works better to do it one of two ways. Either charge your full rate, or do the job for free. (Let them pay for parts of course. They were not free to you.) This way you are not inviting a situation where they still think you should have done it for free, or that you have insulted them with a discounted price they think is still too high.

In some instances you may be able to work out a mutually beneficial barter agreement, say like getting your truck worked on by a mechanic or swapping for other free or discounted goods or services. That of course is between you and your accountant.

M. Scott Gregg

Sometimes it is a good deal and helps everyone, especially at tax time!

Closing Thoughts

You now have a great idea of how you can make way more money in the plumbing and heating field than the average plumber. Hopefully by reading this book you are already thinking of many more ideas that you and other plumbers can use to grow your businesses.

You are now better prepared to deal with your customers. You'll be able to get them to communicate their specific needs and you'll be prepared to meet those needs. You will earn their trust. They will gladly refer you to their family, friends and neighbors

If you're a fleet contractor and use the suggestions in this book, your company will most likely get a reputation for being one of the best and most in-demand contractors in your area. With the increased business will come the increased profit.

It is up to all of us to raise the bar for our trade. It is our responsibility to help our trade get the reputation of professionalism that it deserves.

Please consider the suggested reading section next and keep learning. The more you know and the more you can do, the more you are worth to your customers.

M. Scott Gregg

You will probably think of many other things that could be added to this book and I would like to hear them. If you would like to share any of your good ideas with me (Or even some of the bad ones) I would like get them. Please feel free to e-mail them to me at msgjhg@comcast.net or contact me through this book's publisher.

A Tribute to My Dad

It's because of my dad that this book made it from the idea stage to the computer and then into so many hands, hopefully with the result of enriching their lives.

My dad, Jay W. Gregg, was only fifty-seven when he died from cancer of the esophagus. We all know life is short, Sometimes you truly learn how short it can be.

It's too short not to enjoy every day, including your time at work. It's too short to put off enjoying your time away from work also. Let your work provide you with the lifestyle you want. Take the opportunity to try new things and explore every opportunity.

Suggested Reading:

Who Moved My Cheese?

You really need to read this book! This is a very small book that looks on the outside like a children's book. It takes about an hour to read from cover to cover and will change your life. It will be the best hour you ever spend. This is the book that will show you who you are, why you are where you are in your life and teach you how to deal with your life's changes in a positive way. Through the four main characters in the book you will see yourself, and learn how to be the person you want to be. Do not miss this one. Go buy it as soon as you put this down!

Anything by Zig Ziglar!

Mr. Ziglar is a leader in the field of motivational speakers. His seminars sell out and his books are on bestseller list. When you read them you are motivated to do better things with your time, both to build wealth and to build yourself.

How to win friends and influence people
By Dale Carnagie

The title of this book is misleading in a way. To me it sounds as if it will teach you to deceive people. This book teaches a communication style which will help you get people to reach out to you and will allow you to establish mutually satisfying relationships with them. This is a must read for anyone in any line of work. It will help you deal with all of the people in your life.

9 steps to Financial Freedom

By Suse Orman

This book will show you how to build lasting wealth in your life. It helps you discover ways to handle your money responsibly so that you can keep what you earn and build a legacy for your children. Regardless of how you make your money or how much money you make, this well written book is easy to read and gets you on the right track towards financial freedom.

One Minute Millionaire

By Mark Victor Hansen and Robert G. Allen

This book is far from being the too good to be true thing its title represents. It is an inspirational way to show you the many routes you can take to becoming a real Millionaire! It also helps you learn how to go about it the right way. It is really two books in one. They are written side by side. The first is the non-fiction "Millionaire" part and the second is a fictional book written about a mother who must make a Million dollars in 90 days to get her kids back from her evil, rich in-laws.

Proverbs:

This book in the Bible is the one that is full of inspirational guidance for business. As stated in the foreword of this book, the intention here is not to proselytize. The reason I chose to include these passages is that a large portion of Proverbs gives clear advice on dealing with people and conducting business and that *is* the purpose of this book. It speaks of honesty, learning, discipline, motivation, love, kindness, and many other things that will help you to be a better businessperson overall. Of all of the books in the bible this one is perhaps the easiest to understand. King Solomon, considered by many to be the wisest man to ever live, wrote most of it.

The "Dummies" series of books

All of these books are written by experts in their fields. They are interesting, informative, humorous and entertaining, never boring. They deliver, in an easily understandable manner, a lot of important information, much of it difficult. By the time you finish a "Dummies" book you will have gained the knowledge essential to your field. The books are well worth the price. It would be impossible to list all of the titles in the Dummies series in this space. Chances are if you have an interest in something, there is a Dummies book available that will help you learn more about it. Here are a few that I have used and found helpful

Act! For Dummies
Quickbooks-All in One Desk Reference For Dummies
Excel 2002 For Dummies

There's this housewife who has a talking parrot. She decides to go shopping. On the way out she says goodbye to the parrot, who is sitting on a perch by the door. The parrot, which is just learning to speak, says nothing.

A little while later a plumber comes by and rings the doorbell. Having heard the bell many times before, and having learned the response, the parrot says, "Who is it?"

The plumber says, "It's the plumber. I've come to fix the sink."

No reply so the plumber rings the bell again.

"Who is it?" squawks the bird.

"It's the Plumber. I've come to fix the sink."

Still no reply so he rings the bell again.

"Who is it?" squawks the bird again.

The plumber, clearly getting agitated, raises his voice, **"It's the Plumber. I've come to fix the sink!"**

The poor plumber, exasperated, falls dead of a heart attack at the door. A short time later the lady of the house comes home and steps over the plumber into the house. "I wonder who that is?" she says out loud.

The parrot says "It's the plumber, he's come to fix the sink."

Acknowledgments

Thanks to plbg.com for their contributions to the trade

Thank you to Vic of plumbing supply.com. Many of the quotations in this book are from their site and they are a sponsor of plbg.com

I would like to especially thank my editor Martha Hyams, of North Truro, Massachusetts who was a joy to work with and helped me greatly by putting the "polish" on this book. Thanks also to my Aunt Joyce "Jo-Jo" Garner for getting us together.

About the Author

Scott Gregg is a licensed Master Plumber in the state of Virginia. He began in high school when he found out that low level color blindness would keep him from his lifelong dream of flying fighters for the airforce. While wondering what career to choose, his mother (after paying a plumber for a repair) suggested that he look into that trade as "plumbers always make good money."

He attended Chesterfield Technical Center for his junior and senior year. For the next two years he went to Richmond Technical Center at night, while completing the plumbing apprenticeship program in the field and getting his journeymen's license.

Scott then began work as a Journeyman Plumber and eventually ran projects as a foreman. His first taste of estimating projects came while he was working for Tyler Mechanical.

In the late 80's, Scott found himself back in the field and eventually at the door of Harris Heating and Plumbing, looking for work. When Mr. Harris found out that Scott had some estimating experience he put him to work with Lewis Pickett, their Chief Estimator, as a "temporary" position to help him get caught up. This temporary position lasted almost 7 years.

Scott learned from Lewis Pickett the ropes of dealing with everyone from engineers, field personnel, architects, other estimators and project manager. Soon he became a project manager himself.

When Scott found out about an opportunity to work for a Manufacturer's Representative called Power and Heat Systems, owned by Charlie Hunt, he and the position. He became a salesman for the company. He enjoyed his new position and working with the company. He learned first hand how to deal with the Manufacturers and Engineers at a new level. His experience in the trade gave him the advantage of knowing the needs of his customers from their perspective.

About 4 years later Scott was approached by a large power plant company called The Sam English Company. This company dealt with large industrial boiler projects. They were looking for someone to train as their project manager.

Scott gained a lot of knowledge about large boilers from working on one project that had a boiler built on site that was over seven stories high! He gained hands on experience working with these huge machines.

Eventually the company decided to supplement their large boiler work by branching into bidding some mechanical projects like chiller and cooling tower replacements, mainly due to Scott's experience in bidding and running these types of projects. This quickly proved to be profitable for them.

When the partners found an opportunity to buy a mechanical business they jumped on it and made Scott the project manager for the new company.

During this time, Scott began to get more and more requests from other people, outside of work, to do plumbing repair for them. It started out with a real estate agent and from there it grew.

Before long, Scott bought a work van at an auction and outfitted it for his side work. He incorporated his business and obtained licensing and insurance and continued to build his business to the point where he was making almost as much money working a few hours a day after working the day job as he was as a project manager. Scott Gregg Plumbing was born.

Scott learned how to deal with homeowners and was able to use some of his sales skills by finding out what the customers wanted or needed while he was there. He learned how to ask the right questions, and to listen to the answers. Word of mouth quickly spread to keep Scott as busy as he wanted with high quality, well paying customers.

After a couple of years in business an opportunity came along to get back into the Manufacturer's Rep business again and Scott took the position opening the State of VA for the company.

During many training classes held, Scott found that plumbers indeed did not have any training in how to sell their products or services to maximize their potential. He began searching for a book on the subject. He found out there were no books on the subject and decided to write one himself, with the goal of passing along his sales training and experience to the trade.

Printed in the United States
26645LVS00005B/1-93